MOVING PARTICLE SEMI-IMPLICIT METHOD

MOVING PARTICLE SEMI-IMPLICIT METHOD

A Meshfree Particle Method for Fluid Dynamics

SEIICHI KOSHIZUKA
The University of Tokyo, Tokyo, Japan

KAZUYA SHIBATA
The University of Tokyo, Tokyo, Japan

MASAHIRO KONDO
The University of Tokyo, Tokyo, Japan

TAKUYA MATSUNAGA
The University of Tokyo, Tokyo, Japan

ACADEMIC PRESS

An imprint of Elsevier

Academic Press is an imprint of Elsevier
125 London Wall, London EC2Y 5AS, United Kingdom
525 B Street, Suite 1650, San Diego, CA 92101, USA
50 Hampshire Street, 5th Floor, Cambridge, MA 02139, United States
The Boulevard, Langford Lane, Kidlington, Oxford OX5 1GB, United Kingdom

Notices
Knowledge and best practice in this field are constantly changing. As new research and experience broaden
our understanding, changes in research methods, professional practices, or medical treatment may become
necessary.

Practitioners and researchers must always rely on their own experience and knowledge in evaluating and
using any information, methods, compounds, or experiments described herein. In using such information or
methods they should be mindful of their own safety and the safety of others, including parties for whom they
have a professional responsibility.

To the fullest extent of the law, neither the Publisher nor the authors, contributors, or editors, assume any
liability for any injury and/or damage to persons or property as a matter of products liability, negligence or
otherwise, or from any use or operation of any methods, products, instructions, or ideas contained in the
material herein.

British Library Cataloguing-in-Publication Data
A catalogue record for this book is available from the British Library

Library of Congress Cataloging-in-Publication Data
A catalog record for this book is available from the Library of Congress

ISBN: 978-0-12-812779-7

For Information on all Academic Press publications
visit our website at https://www.elsevier.com/books-and-journals

Working together
to grow libraries in
developing countries

www.elsevier.com • www.bookaid.org

Publisher: Matthew Deans
Acquisition Editor: Brain Guerin
Editorial Project Manager: Thomas Van Der Ploeg
Production Project Manager: Kamesh Ramajogi
Cover Designer: Victoria Pearson

Typeset by MPS Limited, Chennai, India

CONTENTS

Preface   *ix*

1. Introduction   1

1.1   Concept of Particle Methods   1
    1.1.1   Lagrangian Description   2
    1.1.2   Meshless Discretization   3
    1.1.3   Continuum Mechanics   5
1.2   MPS Method   11
    1.2.1   Weighted Difference   11
    1.2.2   Particle Interaction Models   12
    1.2.3   Semi-implicit Algorithm   16
    1.2.4   MPS and SPH   18
1.3   Research History of Particle Methods   20
References   23

2. Fundamental of Fluid Simulation by the MPS Method   25

2.1   The Elements of the MPS Method   26
    2.1.1   Setting the Initial Positions of Particles   28
    2.1.2   Setting Initial Velocities of Particles   29
    2.1.3   How to Move Particles   29
    2.1.4   How to Calculate Acceleration of Particles   31
2.2   Basic Theory of the MPS Method   33
    2.2.1   Mass of a Particle   33
    2.2.2   Governing Equations   35
        2.2.2.1   The Navier—Stokes Equations   35
        2.2.2.2   Equation of Continuity   38
        2.2.2.3   Notation by Vectors   39
    2.2.3   Particle Number Density and Weight Function   40
        2.2.3.1   The Standard Particle Number Density n^0   42
        2.2.3.2   Relationship Between Particle Number Density and Fluid Density   43
        2.2.3.3   Example of Calculation   44
        2.2.3.4   The Form of a Weight Function   45
    2.2.4   Approximation of Partial Differential Operators   45
        2.2.4.1   Gradient   46
        2.2.4.2   The Gradient Model of the MPS Method (Nabla Model)   46

	2.2.4.3	The Meaning of Each Parts of the Gradient Model	47
	2.2.4.4	Example of Gradient Calculation	51
	2.2.4.5	Laplacian Operator and Its Uses	53
	2.2.4.6	The Laplacian Model of the MPS Method	53

2.2.5 Semi-implicit Method   57

2.2.5.1 How to Calculate Pressure, and the Necessity
of the Semi-implicit Method   57

2.2.5.2 The Outline of the Semi-implicit Method in the MPS Method   59

2.2.5.3 Details of the Semi-implicit Method of the MPS Method   61

2.2.5.4 Derivation of Pressure Poison Equation of the MPS Method   68

2.2.5.5 How to Calculate the Pressure Poisson Equation   70

2.2.5.6 The Boundary Condition of Pressure   73

2.2.5.7 The Boundary Condition of Velocity   74

2.3 Outline of Simulation Programs   74

2.3.1 Contents of Program   74

2.3.2 How to Compile and Execute the Sample Programs   84

2.3.3 How to Visualize the Simulation Result   85

2.3.4 Functions of the Program   86

2.3.4.1 Libraries and Declarations   86

2.3.4.2 Main Function   88

2.3.4.3 initializeParticlePositionAndVelocity_for 2dim( ) Function   88

2.3.4.4 calculateNZeroAndLambda( ) Function   90

2.3.4.5 weight( ) Function   90

2.3.4.6 mainLoopOfSimulation( ) Function   91

2.3.4.7 calculateGravity Function   93

2.3.4.8 calculateViscosity Function   93

2.3.4.9 moveParticle( ) Function   95

2.3.4.10 calculatePressure( ) Function   95

2.3.4.11 calculateNumberDensity( ) Function   96

2.3.4.12 setBoundaryCondition( ) Function   97

2.3.4.13 setSourceTerm( ) Function   98

2.3.4.14 setMatrix( ) Function   98

2.3.4.15 solveSimultaniousEquationsBy
GaussianElimination( ) Function   99

2.3.4.16 calculatePressureGradient( ) Function   100

2.3.4.17 calculatePressure_forExplicitMPS( ) Function   101

2.3.4.18 calculatePressureGradient_forExplicitMPS( ) Function   102

2.4 Exercise of Simulation   103

2.4.1 Exercises   103

2.5 Hints of Exercises   104

2.6 Frequently Asked Questions   105

2.6.1 What Is the Best Effective Radius of the Interaction Zone?   105
2.6.2 Why Do We Need to Arrange Dummy Wall Particles Behind
Wall Particles?   105
2.6.3 Particles Penetrated a Wall: What Is the Possible Reason?   106
2.6.4 How Do We Set the Time Increment Δt?   106
2.6.5 It Seems Simulation Diverged Because Particles Exploded:
What Is the Reason?   107
2.6.6 Fluid Was Compressed: What Is the Reason?   107
2.6.7 How Can We Add a Function of Inlet or Outlet
Boundary in a Simulation Program?   107
2.6.8 What Is the Most Time-Consuming Part in an MPS Simulation?   108
2.6.9 What Are the Drawbacks and the Strong Points of
the Semi-implicit Method?   108
References   108

3. Extended Algorithms   **111**

3.1 Compressible-Incompressible Unified Algorithm   111
3.2 Explicit Algorithm Using Pseudo-Compressibility   118
3.3 Symplectic Scheme   125
3.4 Arbitrary Lagrangian-Eulerian   134
3.5 Rigid Body Model   140
3.6 Structural Analysis   146
References   150
Further Reading   153

4. Boundary Conditions   **155**

4.1 Introduction   155
4.2 Solid Wall   159
4.2.1 Wall Particle Representation   162
4.2.2 Mirror Particle Representation   168
4.2.3 Distance Function—Based Polygon Representation   176
4.2.4 Boundary Integral—Based Polygon Representation   180
4.3 Free Surface   189
4.3.1 Free-Surface Particle Detection   193
4.3.2 Pressure Calculation   200
4.4 Inlet and Outlet Boundary Modeling   207
References   212

5. Surface Tension Models in Particle Methods   217

5.1   Surface Tension Calculation Using CSF Continuum Equation   218
   5.1.1   CSF-Based Model Proposed by Nomura et al. (2001)   218
   5.1.2   Other Surface Tension Models Based on CSF Equation   222
5.2   Surface Tension Calculation Based on a Pairwise Potential   223
   5.2.1   Potential-Based Model Proposed by Kondo et al. (2007a,b)   224
   5.2.2   Further Improvement of the Potential-Based Approach   227
   5.2.3   Wettability Calculation in the Potential Model   227
5.3   Applications of the Surface Tension Models Using the MPS Method 229
References   231

6. Advanced Techniques   233

6.1   Liquid—Solid Phase Change Model   233
6.2   Gas—Liquid Two-Phase Flow and Phase Change Model   236
6.3   Turbulence   242
6.4   Suppression of Pressure Fluctuations   245
6.5   Higher-Order Schemes   247
6.6   Parallel Computing   250
6.7   Multiresolutions   254
6.8   V&V and Applications   259
   6.8.1   Verification and Validation   259
   6.8.2   Application to Automobile Industry   260
   6.8.3   Application to Chemical Engineering   265
   6.8.4   Application to Metal Engineering   268
   6.8.5   Application to Biomechanics   270
References   273

Index   281

PREFACE

The Finite Volume Method (FVM) has been established in 1980s for simulating fluid dynamics problems (e.g., Rhie-Chow, 1983). Unstructured grids of arbitrary shapes can be used with the techniques of coordinate transformation that is based on differential geometry. Another typical aspect of the FVM is variable arrangement that is based on sophisticated topology. The algorithm for incompressible flow employs implicit pressure correction procedure using the pressure Poisson equation. The FVM is markedly successful to apply to engineering problems. However, it still has limitations derived from the grid. Numerical errors and instabilities take place where the grid is highly distorted. Besides, grid generation needs heavy effort and various knowledge to obtain reasonable results for complicated three-dimensional geometries.

Gridless/particle methods are expected to solve the troubles concerning the grid in the FVM. Particularly, particle methods using Lagrangian description do not need to calculate the convection terms. It is revolutionary in the computational fluid dynamics because discretization of the convection term is likely to cause terrible numerical diffusion and instability, and massive studies have been dedicated to this topic for many years. The particle methods are essentially free from the troubles concerning the convection terms.

Smoothed particle hydrodynamics (SPH) developed by Lucy (1977) and Gingold and Monaghan (1977) has been used for fluid dynamics problems in astrophysics. The SPH method was limited to nonviscous and compressible flow in the early 1990s. The idea of the Moving Particle Semi-implicit (MPS) method was conceived as the extension of the FVM to a particle method; spatial discretization was based on differences and pressure Poisson equation was employed for incompressible viscous flow (Koshizuka et al., 1996; Koshizuka and Oka, 1997). Engineering problems have been solved by the MPS method. Nowadays, many techniques in the MPS and SPH methods are common: semiimplicit algorithm, explicit algorithm using pseudo-compressibility, free surface boundary condition, wall boundary condition, surface tension model, etc. However, the spatial discretizations are basically different and the resultant discretized equations are clearly distinctive.

This book provides the comprehensive knowledge of the MPS method that has been developed to date. Chapter 1, Introduction, is an introduction to the basic concept of the MPS method. Differences between the MPS and SPH methods are explained. The research history of the particle methods for continuum mechanics is summarized. Chapter 2, Fundamental of Fluid Simulation by the MPS Method, provided a detailed explanation of the basic formulation of the MPS method. This chapter is instructive for the beginners of the MPS method. Chapter 3, Extended Algorithms, explains various algorithms except for the basic semiimplicit one. They are two compressible—incompressible unified algorithms considering slight and strong compressibility, an explicit algorithm using pseudo-compressibility, symplectic schemes based on Hamiltonian system, an Arbitrary Lagrangian-Eulerian (ALE) approach, a rigid body model using a quaternion, and an elastic solid model. Chapter 4, Boundary Conditions, shows the treatment of boundaries: such as solid walls, free surfaces and inlet/outlet boundaries. Chapter 5, Surface Tension Models in Particle Methods, is used for surface motion models used in the MPS method: the model using continuum surface force and pair-wise potential force. Various advanced techniques are summarized in Chapter 6, Advanced Techniques: liquid—solid phase change, gas—liquid two-phase flow, a subparticle-scale turbulence model, numerical techniques for suppressing pressure fluctuations, high-order schemes, parallel computing techniques, and multiresolution techniques. Industrial applications are presented as verification and validation (V&V) examples.

Chapter 1, 3, and 6 were mainly written by Koshizuka, Chapter 2 and Section 6.7 were written by Shibata, Matsunaga wrote Chapter 4, and Kondo authored Chapter 5 and Section 3.6.

The development of the MPS method has been stimulated by the studies in the field of physics-based computer graphics. A calculation method of rigid bodies using quaternions is imported from the physics-based computer graphics. Furthermore, high-quality visualization makes the calculation results of the particle method very attractive, particularly complex free surface motion accompanied by splashing. Parallel computing on graphics processing unit has been realized in the MPS method in a relatively early stage due to the interaction with the physics-based computer graphics.

Another important movement influencing on the MPS method is V&V. V&V has been implemented as the technical guidelines to the computer simulation procedure to assure the credibility of the results. This is important to use the computer simulation in the industries. Application of the MPS method to the industrial problems needs to follow the V&V framework. In this book, application examples in the industries are provided in the V&V section in Chapter 6, Advanced Techniques.

In 2011, the Great East Japan Earthquake occurred suddenly. The tsunami killed many people. One of the lessons of this tsunami was that many floating objects were accompanied to destroy the houses and the structures on the coast. The particle methods should be used more for the analysis of tsunami run-up with floating objects. In addition, unbelievable reactor core melting and devastating hydrogen explosions took place in Fukushima Dai-ichi Nuclear Power Plant after the tsunami. The first application of the MPS method was the molten fuel fragmentation process in the postulated severe accident of the nuclear reactor, and the second paper of the MPS method was published in Nuclear Science and Engineering in 1996. It should be said that the particle methods would be used more for nuclear reactor safety. Visualization and V&V are more and more important to transfer the simulation results of the disasters to the public and decision makers.

The authors would like to appreciate all the members who studied or helped studying the MPS method in our laboratories. The authors also would like to express thanks to the researchers and engineers who developed and used the software using the MPS method in the companies. The contents of the book are accumulation of their huge studies and tough experiences. The kind proposal, proper guidance, and encouragement to writing manuscripts by Mr. Brian Guerin, Mr. Thomas van der Ploeg, and Ms. Swapna Srinivasan in Elsevier are highly acknowledged.

Seiichi Koshizuka, Kazuya Shibata,
Masahiro Kondo and Takuya Matsunaga
January 2018

CHAPTER 1

Introduction

Abstract

Particle methods are based on Lagrangian description and meshless discretization to simulate continuum mechanics. Advantages and disadvantages of the particle methods are discussed from these viewpoints. Moving particle semi-implicit (MPS) method is one of the particle methods. The fundamental idea of the MPS method is a weighted difference without the mesh. Particle interaction models are prepared for differential operators and substituted into the governing equations. Pressure Poisson equation is constructed to solve the pressure field implicitly, while the other terms are explicitly calculated. Differences between the MPS method and smoothed particle hydrodynamics (SPH) method are explained. The research history of the particle methods for continuum mechanics is summarized.

Keywords: Lagrangian description; meshless discretization; moving particle semi-implicit (MPS) method; weight function; particle interaction model; semi-implicit algorithm; incompressible flow; pressure Poisson equation

Contents

1.1 Concept of Particle Methods	1
1.1.1 Lagrangian Description	2
1.1.2 Meshless Discretization	3
1.1.3 Continuum Mechanics	5
1.2 MPS Method	11
1.2.1 Weighted Difference	11
1.2.2 Particle Interaction Models	12
1.2.3 Semi-implicit Algorithm	16
1.2.4 MPS and SPH	18
1.3 Research History of Particle Methods	20
References	23

1.1 CONCEPT OF PARTICLE METHODS

The concept of the particle methods is explained. The most important viewpoint is the comparison with the mesh methods (Fig. 1.1). The mesh methods are, for example, the finite volume method and the finite element method, which have been mainly used for the computational fluid dynamics and the computational solid mechanics, respectively.

Moving Particle Semi-implicit Method
DOI: https://doi.org/10.1016/B978-0-12-812779-7.00001-1

1

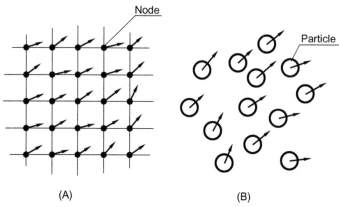

(A)   (B)

Figure 1.1 (A) Mesh method and (B) particle method.

The particle methods are said as new and advanced against the conventional mesh methods. In this section, the differences between particle and mesh methods are summarized to three aspects and both advantages and disadvantages are discussed.

1.1.1 Lagrangian Description

Particles are used in the particle methods. They are the calculation points where the variables are located. The variables are, for instance, velocity vector components, pressure, temperature, etc. The particles are equivalent to the nodes, where the variables are located, in the mesh method.

The particles move with their velocities, while the nodes do not. We can say that the particles are fixed on the moving material. This is expressed as Lagrangian description. On the other hand, the nodes of the mesh methods are fixed in space, which is expressed as Eulerian description.

In the particle methods, it is necessary to hold coordinate vector components as variables as well as the velocity vector components. This means that the additional calculation time is required to update the coordinate vector components in each time step. Actually, it is not very large in the total calculation time.

A moving particle can be regarded as a substance of the material keeping a constant mass. Image of the moving particles is that the divided materials travel with their own velocities. The velocity and coordinate vectors of a particle are considered as those of the center of gravity of the divided material. The distribution of the mass is not considered in the particle. Therefore, overlapping of the particles is not concerned.

Each particle is simply expressed by a velocity vector and a coordinate vector and has a mass without considering the mass distribution.

Mass conservation is essentially satisfied in the particle methods when the number of the initial particles is kept. Addition of new particles means mass increase and deletion of existing particles means mass decrease. This characteristic of the particle method is advantageous in the computational fluid dynamics, particularly, for incompressible fluids. It has been considered that rigorous mass conservation is important for the analysis of incompressible fluids because the pressure field is highly sensitive to the mass conservation. The mesh methods are based on space and the mass conservation is derived by satisfying the continuity equation. The mesh methods rigorously keep the space by explicitly drawing the mesh. Mass is indirectly kept by using the continuity equation which represents the relation between mass and space. The continuity equation expresses change of density, which is mass divided by space. In the particle methods, mass is conserved directly without the continuity equation.

The space that is occupied by a particle cannot be rigorously determined in the particle methods. Mass distribution inside a particle is not considered. A particle simply represents its velocity and coordinate vectors of its center of gravity. Therefore, there are no typical directions or coordinates of the particle. It can be said that the particle is spherical. In the mesh methods, the mesh near the wall is often generated to have a large aspect ratio; the mesh length parallel to the wall is longer than that vertical to the wall. Such "thin" mesh is efficient and widely used in the computational fluid dynamics since the variable distributions change rapidly near the wall. However, the aspect ratio that is different from 1.0 is difficult in the particle methods. This is one of the disadvantages of the particle methods.

We can say that the concept of moving particles, Lagrangian description, has an advantage of essential mass conservation and a disadvantage of difficulty using the technique of the arbitrary aspect ratio.

1.1.2 Meshless Discretization

The particle methods need no mesh for discretization of the governing equations, while the mesh methods are based on the mesh. First, the function of mesh is discussed in the mesh method. As shown in Fig. 1.1, the mesh consists of segments, each of which connects two nodes. Connection means that two nodes are adjacent. Discretized equations are given as the relationships among the adjacent nodes. For example, in the two-dimensional space (Fig. 1.1), one node has four segments which reach four adjacent nodes. A difference scheme is to be constructed using the variables located at five nodes: one node and its four adjacent nodes. We can say that the mesh is used to identify the adjacent nodes which are to be used for discretization of the governing equations. The mesh explicitly shows the adjacency.

There are no such segments in the particle methods. Adjacent particles are not explicitly shown. The adjacent particles are identified in the simulation by evaluating the distance between two particles (Fig. 1.2). We assume that there are two particles i and j and their position vectors are $\mathbf{r}_i$ and $\mathbf{r}_j$, respectively. When the distance $r\ (= |\mathbf{r}_j - \mathbf{r}_i|)$ is smaller than the effective radius r_e, they are considered as adjacent. The adjacent particle j is called a neighboring particle. The distances of all combinations of two particles are evaluated so that the list of neighboring particles is completed. Discretization equations are to be constructed using the variables at the neighboring particles.

Mesh generation is not necessary in the particle methods. Mesh generation is often very complicated and time consuming in three-dimensional complex domains which are solved in the industrial applications. Accuracy of the simulation result is largely affected by the mesh quality. A more accurate result is expected by using a finer mesh though it requires more computation time. Compromise is necessary in mesh generation by considering the accuracy and the computation time. A lot of know-hows are alleged to be utilized, which leads to the mesh generation complicated. In the particle methods, initial arrangement of the particles is necessary but it is much easier than the mesh generation. This is a substantial advantage of the particle methods.

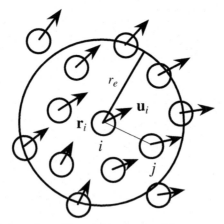

Figure 1.2 Particle interaction with neighboring particles.

In the particle methods, the neighboring particles are memorized as the neighbor list and it is updated in each time step. This is an additional procedure in each time step in contrast to the mesh generation which is carried out once in the initial process. Besides, this procedure is time consuming because the calculation of the distances between all combinations of two particles requires the order of N^2, where N is the total number of particles. The order of calculation is usually N^1 for the substitution of variables into the discretized equations and $N^{1.5}$ for typical solvers of simultaneous linear equations. The order of N^2 has the same order of global interaction physics, e.g., electromagnetic force and gravity, where the interaction force reaches infinity. For the search of the neighboring particles, an algorithm of the order of N^1 has already been proposed and widely used in the particle methods. This algorithm is explained in the later chapters. Thus, at this moment, the additional computation time for making a neighbor list is not a disadvantage of the particle methods any more.

The mesh has a role of dividing space as well as explicit description of adjacency. Consistent division of space ($=$no overlapping) is necessary for consistent integration in the finite element method and for rigorous conservation of physical quantities in the finite volume method. On the other hand, division of space is ambiguous in the particle method. However, distinct troubles or disadvantages are not found due to this at this moment.

1.1.3 Continuum Mechanics

The governing equations for continuum mechanics to be solved are mass and momentum conservation equations. In fluid dynamics, they are also called continuity and Navier–Stokes equations, respectively. The governing equations for continuum mechanics should be discretized without the help of mesh in the particle methods. As explained in Section 1.1.2, mesh has information of strict space division, which cannot be utilized in the particle methods. We need other approaches for the spatial discretization.

For example, in the moving particle semi-implicit (MPS) method (Koshizuka and Oka, 1996), weighted average of differences is the basic concept of the discretization. In the smoothed particle hydrodynamics (SPH) method (Lucy, 1977; Gingold and Monaghan, 1977), superposition of kernels is basically assumed to obtain the spatial distribution of the variables. In both of the particle methods, the mesh is not used for the discretization at all. The detailed formulation of the discretization in the MPS method is explained in Chapter 2, Fundamental of Fluid Simulation by the MPS Method.

The coordinate and velocity vectors of the particles are updated in each time step by using the discretized governing equations. The discretized governing equations involve the variables of the neighboring particles. This can be said that the particle motion is determined through the interaction with the neighboring particles. The discretized governing equations are regarded as the motion equations of the particles. Here, it should be noted that the particles are artificial for discretization in the particle methods. The fluid is represented by a finite number of artificial particles. They are not real particles, such as powder and molecules.

An example of the real particle dynamics is bead motion (Fig. 1.3). We can see the motion of beads when they are poured into a glass as shown in Fig. 1.3A. The bead dynamics can be described by Newton's second law with friction terms. On the other hand, we can see motion of water when it is poured into the same glass as shown in Fig. 1.3B. The governing equations of water are the Navier–Stokes equations. The water behavior is different from that of the beads because the governing equations are different. However, as shown in Fig. 1.3C, the particle

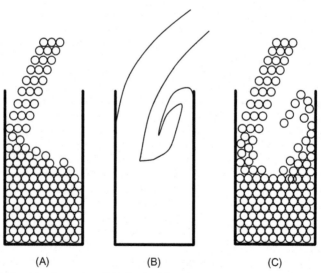

Figure 1.3 Particle dynamics: (A) physical bead motion, (B) physical water motion, and (C) numerical water motion using particle method.

motion can be evaluated based on the discretized Navier—Stokes equations (=particle motion equations) in a computer. In this case, the particle motion in the computer is like water motion. This is the concept of the particle methods using artificial particles for continuum mechanics.

Some other particle methods, molecular dynamics and discrete element method, are the numerical methods for the real particles. In molecular dynamics, motion of molecules is simulated. In discrete element method, motion of powder particles is simulated. These particle methods are essentially different from the present particle methods, such as MPS and SPH, for the continuum mechanics.

Here, a one-dimensional pure convection problem is discussed to understand the difference between the mesh methods and the particle methods. Pure convection of a passive scalar variable ϕ in one-dimensional space x with a constant velocity u (>0) is governed by the following particle differential equation:

$$\frac{\partial \phi}{\partial t} = -u\frac{\partial \phi}{\partial x} \qquad (1.1)$$

The passive scalar variable can be temperature, chemical spices concentration, etc. The pure convection governed in Eq. (1.1) is simple movement of the variable without any change of the profile. An analytical solution is illustrated in Fig. 1.4.

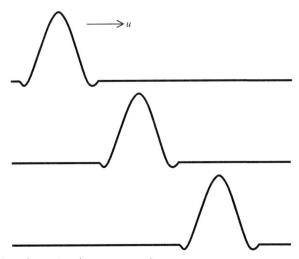

Figure 1.4 One-dimensional pure convection.

First, we consider the mesh method (finite difference method) employing the Eulerian description. The one-dimensional space is discretized to a finite number of mesh points. Each point i holds its coordinate x_i and variable ϕ_i. Temporal discretization is also carried out and the corresponding indices are shown by a superscript as x_i^k and ϕ_i^k at time step k.

First, discretization in time is carried out. The left-hand side of Eq. (1.1) is then discretized to

$$\frac{\partial \phi}{\partial t} = \frac{\phi^{k+1} - \phi^k}{\Delta t} \tag{1.2}$$

where Δt is the time interval between time steps k and $k + 1$. Unknown variables at new time step $k + 1$ are calculated by using known variables at old time step k:

$$\phi_i^{k+1} = \phi_i^k - \Delta t u \frac{\partial \phi}{\partial x}\bigg|_i^k \tag{1.3}$$

The scalar variables at time step $k + 1$ are calculated using Eq. (1.3) which is the discretized equation of Eq. (1.1). The coordinate of each mesh point is not changed because of the Eulerian description. This can be written as:

$$x_i^{k+1} = x_i^k \tag{1.4}$$

Next, we need to discretize $\frac{\partial \phi}{\partial x}\big|_i^k$ in Eq. (1.3) in space. Various spatial difference schemes have been proposed until now. Here, an upwind scheme, one of the basic schemes, is applied:

$$\frac{\partial \phi}{\partial x}\bigg|_i^k = \frac{\phi_i^k - \phi_{i-1}^k}{\Delta x} \tag{1.5}$$

Applying Eq. (1.5) to Eq. (1.3) results in Fig. 1.5A. We can see diffusive profiles of ϕ. The analytical solution of Eq. (1.1) is the simple translational movement of the initial profile with velocity u. The difference between the numerical result and the analytical solution is substantial.

When the central difference scheme, another basic scheme,

$$\frac{\partial \phi}{\partial x}\bigg|_i^k = \frac{\phi_{i+1}^k - \phi_{i-1}^k}{2\,\Delta x} \tag{1.6}$$

is applied, the result is Fig. 1.5B. We can see oscillatory profiles. It is well known that the finite difference method causes substantial numerical errors, which can be diffusive or oscillatory, for convection.

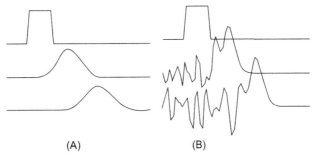

Figure 1.5 Finite difference method for one-dimensional pure convection: (A) upwind difference scheme and (B) central difference scheme.

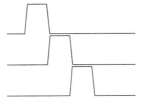

Figure 1.6 Particle method for one-dimensional pure convection.

In the Lagrangian description, the one-dimensional pure convection problem is written as:

$$\frac{D\phi}{Dt} = \frac{\partial \phi}{\partial t} + u\frac{\partial \phi}{\partial x} = 0 \tag{1.7}$$

using the Lagrangian time derivative $\frac{D}{Dt}$. Eq. (1.7) is equivalent to Eq. (1.1). The one-dimensional space is discretized to a finite number of particles. Each particle i holds its coordinate x_i and variable ϕ_i. Unknown variables at new time step $k + 1$ are calculated by using known variables at old time step k:

$$\phi_i^{k+1} = \phi_i^k \tag{1.8}$$

$$x_i^{k+1} = x_i^k + \Delta tu \tag{1.9}$$

The scalar variables at new time step $k + 1$ are the same as those at old time step k as shown in Eq. (1.8) because the particles move with holding their variables. On the other hand, the coordinates of the particles at new time step are changed by adding Δtu as shown in Eq. (1.9). Here, no diffusive or oscillatory errors take place as shown in Fig. 1.6. This is because the convection is simply represented by the Lagrangian motion of the particles.

The extremely accurate calculation for the pure convection in the Lagrangian description must have been noticed by researchers because Eq. (1.7) is well known in fluid dynamics. This characteristic is important in many problems, e.g., flow of a temperature-dependent fluid. When the fluid properties, such as viscosity, are dependent on temperature, accurate calculation of the temperature profile is primarily important. Numerical diffusion or oscillation degrades the accuracy so much. If the mesh methods would be used in the Lagrangian description, mesh tangling might be inevitable and re-meshing would be required. Re-meshing also causes numerical errors because both the re-meshing and the convection require essentially the same calculation; the viewpoint is moved on the material. The particle methods without the mesh lead to remarkably accurate calculation in the pure convection problem using the Lagrangian description.

However, we have to be careful that the particle methods need to calculate the motion of particles for the convection. This makes another type of errors; for instance, simple rotation of a particle makes its rotating radius larger as shown in Fig. 1.7. Higher-order discretization of the particle motion to reduce this type of errors remains for future studies.

It should be noted that the viewpoint in the Lagrangian description is fixed at the fluid only at this viewpoint. The coordinate system is given

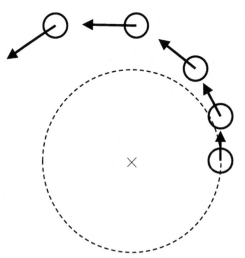

Figure 1.7 Errors in simple rotation using particle method.

in space and not affected by the fluid motion. In other words, x-direction is defined in space and not changed even if the fluid is rotating. This situation is different from the coordinate system fixed on the material, where x-direction rotates as the fluid does. Centrifugal and Coriolis forces emerge when the coordinate system rotates in space. Such additional forces are not necessary to consider in the Lagrangian description.

1.2 MPS METHOD

In this book, the MPS method (Koshizuka and Oka, 1996) is mainly explained. The MPS method has been developed to analyze incompressible flows, such as water and oil. In reality, any fluids are compressible. Incompressibility is an approximation where the sound speed in the fluid is infinity. Most of the fluid flows in the engineering applications are considered as incompressible, since the sound speed is much higher than the flow speed. In this section, the basic concept of the MPS method is introduced. The details are explained in later chapters.

1.2.1 Weighted Difference

A difference can be calculated between two particles as shown in Fig. 1.8. The mesh is not necessary for this difference. The calculation result appears between two particles; the difference is obtained on the segment, the edges of which are two particles. This is the basic concept of spatial discretization of the MPS method.

The difference can be calculated with an arbitrary pair of the particles, but we restrict the calculation within the neighborhood. This is expressed

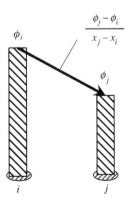

Figure 1.8 Difference between two particles.

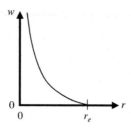

Figure 1.9 Weight function.

by introducing an effective radius r_e. When the distance r between two particles is less than r_e, the difference is calculated. Otherwise, the difference is not calculated.

The first-order difference at particle i in one-dimensional space is calculated as a weighted average of the differences with its neighboring particles:

$$\left\langle \frac{d\phi}{dx} \right\rangle_i = \frac{1}{n_i} \sum_{j \neq i} \frac{\phi_j - \phi_i}{x_j - x_i} w\left(\left|x_j - x_i\right|\right) \tag{1.10}$$

where $w(r)$ is a weight function:

$$w(r) = \begin{cases} \dfrac{r_e}{r} - 1 & r \leq r_e \\ 0 & r > r_e \end{cases} \tag{1.11}$$

and n_i is the summation of the weight function values:

$$n_i = \sum_{j \neq i} w\left(\left|x_j - x_i\right|\right) \tag{1.12}$$

Eq. (1.11) is a typical weight function used in the MPS method as shown in Fig. 1.9. The weight function w decreases with r and $w = 0$ at $r \geq r_e$. The weight function w is infinite at $r = 0$, which actually does not appear in Eq. (1.10) because the pair of particles are made between particle i and its neighbors j. In addition, the infinite value at $r = 0$ is preferable in order to avoid clustering of the particles through the incompressibility condition as explained in Section 1.2.3.

1.2.2 Particle Interaction Models

In the MPS method, particle interaction models are prepared for differential operators in vector calculus, for instance, gradient and Laplacian. This is like the finite volume method where the special discretization formulations are given based on the staggered variable arrangement. For instance, the divergence of the velocity in a cell is formulated as the inflows and outflows of the velocity vector components located on its cell faces.

In three dimensions, the gradient vector, one of the first-order differentiations in vector calculus, is represented by the following discretized equation at particle i in the MPS method:

$$\langle \nabla \phi \rangle_i = \frac{d}{n_i} \sum_{j \neq i} \frac{\phi_j - \phi_i}{|\mathbf{r}_j - \mathbf{r}_i|} \frac{\mathbf{r}_j - \mathbf{r}_i}{|\mathbf{r}_j - \mathbf{r}_i|} w\left(|\mathbf{r}_j - \mathbf{r}_i|\right) \tag{1.13}$$

where $\mathbf{r}_i = (x_i, y_i, z_i)$ is the position vector of particle i. $\frac{\phi_j - \phi_i}{|\mathbf{r}_j - \mathbf{r}_i|}$ is the absolute value of the gradient vector between two particles i and j, and $\frac{\mathbf{r}_j - \mathbf{r}_i}{|\mathbf{r}_j - \mathbf{r}_i|}$ is the unit vector from particle i to particle j. Quantity d is the number of spatial dimensions: $d = 2$ for two dimensions and $d = 3$ for three dimensions. $\frac{\phi_j - \phi_i}{|\mathbf{r}_j - \mathbf{r}_i|} \frac{\mathbf{r}_j - \mathbf{r}_i}{|\mathbf{r}_j - \mathbf{r}_i|}$ stands for the gradient vector between particles i and j. The right-hand side of Eq. (1.13) is the weighted average of the gradient vectors between particle i and its neighboring particles j (Fig. 1.10). This is calculated by multiplying the weight function w, taking sum with respect to j for all particles except for i, and divided by n_i which is the sum of w as shown in Eq. (1.12). The number of spatial dimensions is

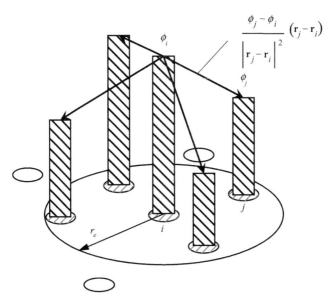

Figure 1.10 Gradient model.

multiplied since the gradient vector calculated between two particles has only one-dimensional information in the direction of $\frac{\mathbf{r}_j - \mathbf{r}_i}{|\mathbf{r}_j - \mathbf{r}_i|}$. This is a half in two dimensions and one-third in three dimensions.

The Laplacian operator, which is the second-order spatial differentiation and the combination of a divergence operator and a gradient operator, is given by:

$$\langle \nabla^2 \phi \rangle_i = \frac{2d}{\lambda_i n_i} \sum_{j \neq i} \left(\phi_j - \phi_i \right) w\left(|\mathbf{r}_j - \mathbf{r}_i| \right) \tag{1.14}$$

where

$$\lambda_i = \frac{\sum_{j \neq i} |\mathbf{r}_j - \mathbf{r}_i|^2 w\left(|\mathbf{r}_j - \mathbf{r}_i| \right)}{\sum_{j \neq i} w\left(|\mathbf{r}_j - \mathbf{r}_i| \right)} \tag{1.15}$$

Eq. (1.14) means that a part of the quantity of particle i is transferred to particle j

$$\langle \Delta \phi_{i \rightarrow j} \rangle_i = \frac{2d}{\lambda_i n_i} \phi_i w\left(|\mathbf{r}_j - \mathbf{r}_i| \right) \tag{1.16}$$

and simultaneously a part of the quantity of particle j is transferred to particle i,

$$\langle \Delta \phi_{j \rightarrow i} \rangle_i = \frac{2d}{\lambda_i n_i} \phi_j w\left(|\mathbf{r}_j - \mathbf{r}_i| \right) \tag{1.17}$$

where $\langle \rangle_i$ means the evaluation at particle i. The quantity represented by Eq. (1.16) is lost from particle i and that represented by Eq. (1.17) is obtained by particle i as shown in the right-hand side of Eq. (1.14).

The quantity obtained by particle j is

$$\langle \Delta \phi_{i \rightarrow j} \rangle_j = \frac{2d}{\lambda_j n_j} \phi_i w\left(|\mathbf{r}_i - \mathbf{r}_j| \right) \tag{1.18}$$

If λ_i and n_i keep the same values for all particles, i.e., $\lambda_i = \lambda_j$ and $n_i = n_j$, Eqs. (1.16) and (1.18) are the same and the quantity lost from particle i is just obtained by j. In this case, the quantity is locally conserved between particles i and j in the particle interaction model with respect to the Laplacian operator.

The Laplacian operator represents diffusion in physics. The discretized formulation of the Laplacian operator in the MPS method represents distribution of the quantity from particle i to its neighboring particles j

(Fig. 1.11). The distribution is based on the weight function and the mesh is not necessary. This discretized formulation is natural.

In the time-dependent diffusion process, quantity ϕ obeys the following governing equation:

$$\frac{\partial \phi}{\partial t} = D\nabla^2 \phi \tag{1.19}$$

where D is diffusivity. At $t = 0$, we assume that particles i have values of ϕ_i, which is like the superposition of delta functions:

$$\phi(\mathbf{r}, t = 0) = \sum_i \phi_i \delta(\mathbf{r}_i) \tag{1.20}$$

At time t, the analytical solution of the superposition of the delta functions is the superposition of Gaussian functions:

$$\phi(\mathbf{r}, t) = \sum_i \phi_i \left(\frac{1}{\sqrt{4\pi Dt}}\right)^d \exp\left(-\frac{|\mathbf{r} - \mathbf{r}_i|^2}{4Dt}\right) \tag{1.21}$$

because the analytical solution of a delta function is a Gaussian function. If we employ the Gaussian function as the weight function, the analytical solution is obtained by using the particle interaction model of the Laplacian operator. The Gaussian function would be better than the weight function of the MPS method, Eq. (1.11). However, the

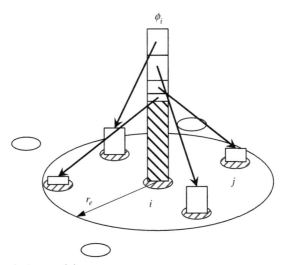

Figure 1.11 Laplacian model.

Gaussian function has the following two problems as the weight function in the particle methods. The first one is that the Gaussian function has the infinite range of interaction. This requires much calculation time and it is not acceptable in large-scale problems. The second one is that the Gaussian function is not accurate in the discretized calculation. The Gaussian function decreases rapidly with the increase in the particle distance $|\mathbf{r}|$. If we would repeatedly apply the Gaussian function in each time step in the particle methods, the values at only the particle positions would be used for the next superposition at the next time step. Actually, there are no neighboring particles where the Gaussian function changes rapidly. A more flat distribution of the weight function, like Eq. (1.11), is better for this repeated superposition at the discretized particle positions.

It is known that the repeated superposition of a Gaussian function in the continuous field is also a Gaussian function. The repeated superposition of any function converges to the Gaussian function if the variance increase is the same as that of the Gaussian function. This is known as the central limit theorem. When any function is applied to the Laplacian model of Eq. (1.14) for the discretized calculation of Eq. (1.19), the variance increase is $2dDt$, which is the same as that of the analytical solution. The Laplacian model of the MPS method keeps the variance increase of $2dDt$ by introducing λ_i and n_i. Thus, Eq. (1.11) is preferred to the Gaussian function for the particle interaction model of the Laplacian operator.

1.2.3 Semi-implicit Algorithm

The MPS method was developed for incompressible flow analysis with free surfaces (Koshizuka and Oka, 1996). The governing equations of incompressible flows are the continuity equation and the Navier—Stokes equations. The continuity equation can be written as:

$$\nabla \cdot \mathbf{u} = 0 \qquad (1.22)$$

or

$$\frac{D\rho}{Dt} = 0 \qquad (1.23)$$

where ρ and $\mathbf{u}$ are density and velocity vector, respectively. $\frac{D}{Dt}$ is the Lagrangian time derivative. In the MPS method, Eq. (1.23) is used, while

Eq. (1.22) is usually used in the finite volume method. The Navier–Stokes equations are written as:

$$\frac{D\mathbf{u}}{Dt} = -\frac{1}{\rho}\nabla P + \frac{\mu}{\rho}\nabla^2\mathbf{u} + \frac{\mathbf{f}}{\rho} \qquad (1.24)$$

where P, μ, and $\mathbf{f}$ are pressure, viscosity, and external force vector, respectively.

In the MPS method, the continuity equation and the pressure gradient term in the Navier–Stokes equations are evaluated at the new time step and the other terms in the Navier–Stokes equations are evaluated at the old time step:

$$\left.\frac{D\rho}{Dt}\right|^{k+1} = 0 \qquad (1.25)$$

$$\frac{D\mathbf{u}}{Dt} = -\frac{1}{\rho}\nabla P\Big|^{k+1} + \frac{\mu}{\rho}\nabla^2\mathbf{u}\Big|^{k} + \frac{\mathbf{f}}{\rho}\Big|^{k} \qquad (1.26)$$

where superscripts k and $k+1$ represent the old and new time steps, respectively.

In the MPS method, Eqs. (1.25) and (1.26) are solved by using two steps. In the first step, the viscosity and external force terms are calculated at the old time step and temporary velocity vector is obtained:

$$\mathbf{u}^* = \mathbf{u}^k + \Delta t\left[\frac{\mu}{\rho}\nabla^2\mathbf{u} + \frac{\mathbf{f}}{\rho}\right]^k \qquad (1.27)$$

where superscript $*$ stands for the temporary value. Since the variables at the old time step are known, Eq. (1.27) can be calculated by substitution. This is called an explicit step.

In the second step, the following pressure Poisson equation

$$\nabla^2 P^{k+1} = -\frac{\rho}{\Delta t^2}\frac{n^* - n^0}{n^0} \qquad (1.28)$$

is solved and the new time pressure is obtained. Here, n^* and n^0 are the particle number densities, as defined in Eq. (1.12), after the first explicit step and in the initial condition, respectively.

The temporary velocity vector is revised by evaluating the pressure gradient term at the new time step:

$$\mathbf{u}^{k+1} = \mathbf{u}^* - \frac{\Delta t}{\rho}\nabla P\Big|^{k+1} \qquad (1.29)$$

In Eq. (1.28), the left-hand side is discretized by the Laplacian model (Eq. 1.14) and we have simultaneous linear equations with respect to the new time pressure variables. Since the variables involved in each equation are multiple, the simultaneous linear equations are solved together by a solver, such as the conjugate gradient method. This is called an implicit step. The present semi-implicit algorithm consists of the first explicit step and the second implicit step. The semi-implicit algorithm had been widely used in the finite volume method for incompressible flows and was introduced to the MPS method.

However, there is a distinct difference in the formulations between the MPS method and the finite volume method. The source term of the pressure Poisson equation is deviation of the temporary particle number density n^* from the initial value n^0 in the MPS method. The particle number density is proportional to the weight density ρ. Thus, the source term means the deviation of the weight density from the constant value. The pressure field is implicitly evaluated to keep the weight density constant, which means that Eq. (1.23) should be satisfied as the continuity equation. On the other hand, Eq. (1.22) is used in the finite volume method so that the source term of the pressure Poisson equation is represented by the divergence of the velocity.

The direct requirement for the incompressibility condition is to satisfy Eq. (1.23). If a substantial error remains in the solver of the pressure Poisson equation, the error is carried over in the next time step. The right-hand side of Eq. (1.28) is not zero unless the particle number density is the same as the initial particle number density, and the pressure field is evaluated again in the nest time step. This is good to avoid the accumulation of the errors of the incompressibility condition. In addition, Eq. (1.28) is effective for uniform distribution of the particles. The boundary condition of free surfaces is simple and robust when Eq. (1.28) is used. It can be said that the pressure Poisson equation of Eq. (1.28) is the key point of the success of MPS method for complicated free surface flows.

1.2.4 MPS and SPH

SPH is another particle method for continuum mechanics. The SPH method was initially developed using a fully explicit algorithm for compressible nonviscous flows and mainly applied to astrophysics where the density changes so much ranging from the inside of stars to vacuum. To date, a semi-implicit algorithm has also been introduced to SPH and called ISPH (incompressible SPH). A fully explicit algorithm is also used in the MPS method and called EMPS (explicit MPS). Numerical techniques for the wall boundary conditions, surface tensions, phase changes, etc. are almost common between SPH and MPS.

Nevertheless, there is a significant discrepancy between MPS and SPH. It is the formulation of the spatial discretization. The spatial discretization in the MPS method is based on a difference as explained in Section 1.2.1. In the SPH method, superposition of kernels is first performed to construct the global distribution of a variable (Fig. 1.12) (Monaghan, 1988):

$$\langle \phi(\mathbf{r}) \rangle = \sum_j \frac{m_j}{\rho_j} \phi_j w \left(\left| \mathbf{r} - \mathbf{r}_j \right| \right) \tag{1.30}$$

where m_j and ϕ_j are the mass and the quantity of particle j, respectively. Density ρ_i at the position of particle i is

$$\rho_i = \langle \rho(\mathbf{r}_i) \rangle = \sum_j m_j w \left(\left| \mathbf{r}_i - \mathbf{r}_j \right| \right) \tag{1.31}$$

Here, w is called kernel in SPH. This is almost equivalent to the weight function in the MPS method.

By differentiating the global distribution (Eq. 1.30), the discretized formulation of the differentiation is deduced:

$$\langle \nabla \phi(\mathbf{r}) \rangle = \sum_j \frac{m_j}{\rho_j} \phi_j \nabla w \left(\left| \mathbf{r} - \mathbf{r}_j \right| \right) \tag{1.32}$$

It is superposition of the differentiations of the kernels.

We can see two large discrepancies between Eq. (1.32) in the SPH method and Eq. (1.13) in the MPS method. One is that the differentiation of the kernel appears in the right-hand side of Eq. (1.32), while the differentiation of the weight function does not appear in the right-hand

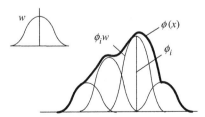

Figure 1.12 Superposition of kernels in SPH.

side of Eq. (1.13). In the MPS method, we do not need to care for the differentiation of the weight function.

The other discrepancy is that the quantity can be given at any positions in space in the SPH method, while it is only given at the particle positions in the MPS method. $\langle \nabla \phi \rangle$ is a function of $\mathbf{r}$ in the SPH method as shown in Eq. (1.32). On the other hand, $\langle \nabla \phi \rangle$ is evaluated at particle i as shown in Eq. (1.13).

If the spatial integration would be required, the global distribution in SPH might be preferable. Actually, the governing equations are discretized to the motion equations of particles where the values at the particle positions are necessary and the global distribution is not used.

In the MPS method, the effective radius r_e, which is a parameter to define the size of the weight function, is different in operators; e.g., $r_e = 2.0 l_0$ and $r_e = 3.1 l_0$ are used for the particle interaction models of the gradient and Laplacian operators, respectively, where l_0 represents the spacing between adjacent particles. In the SPH method, the kernel size is given when the global distribution is assumed. To be consistent with this global distribution, we cannot change the kernel size when differentiating the distribution.

Corresponding to these discrepancies, the SPH method may have a little different concept from the MPS method. In SPH, the kernel is considered as a mass distribution of each particle. The superposition of the kernels represents the physical superposition of mass. The kernel has the largest value at the particle center and it decreases as the distance from the particle center increases. Thus, the particle is like a spherical cloud. This concept may be more fitted to compressible fluids.

1.3 RESEARCH HISTORY OF PARTICLE METHODS

Particle-and-force (PAF) method is recognized as the first particle method for fluid dynamics (Harlow and Meixner, 1961; Harlow, 1963; Daly et al., 1965). The fluid flows were solved by the motion of particles which interact each other. PAF was originated from the particle-in-cell (PIC) method, which used both particles and a mesh. PAF was developed by eliminating the mesh from PIC (Harlow, 1988). There were two terms for particle interaction in the governing equation of PAF; one was based on the equation of state and the other was introduced as an artificial viscosity. The artificial viscosity was also employed in SPH later. The advantages and disadvantages of the Lagrangian mesh method, the

Eulerian mesh method, PIC, and PAF were summarized in a table by Daly et al. (1965). It is marvelous that these advantages and disadvantages are reasonable even when viewed from today. However, only a few calculation examples were shown in the papers of PAF.

SPH was proposed in 1977 (Lucy, 1977; Gingold and Monaghan, 1977). Astrophysical problems were solved as compressible inviscid flows. A full explicit algorithm was used. For compressible flows, the particle method worked as an adaptive method. The particles were clustered where the density was high and high spatial resolution was necessary. The particles were dispersed where the density was low. This characteristic might be fitted to the astrophysical problems.

Water flows with free surfaces were solved by using the governing equation of compressible flows with an artificial viscosity (Monaghan, 1994). The fully explicit algorithm was kept. The equation of state was employed with assuming an artificially smaller sound speed. This algorithm is now called weakly compressible SPH (WCSPH). The artificially smaller sound speed made the numerical calculations more stable and we could use a larger time step.

A semi-implicit algorithm was developed for the particle method in the MPS method (Koshizuka and Oka, 1996) for incompressible flows with free surfaces. An implicit pressure Poisson equation was formulated to keep the particle number density constant. The semi-implicit algorithm had been widely used in the finite volume methods. The difference is the source term of the pressure Poisson equation; the particle number density is used in the MPS method, while the velocity divergence is used in the finite volume method. The usage of the particle number density, which is proportional to the weight density, is straightforward for the incompressibility condition. In addition, the distribution of particles is kept uniform after solving the pressure Poisson equation. However, this source term caused large fluctuations of the pressure field in both space and time. The source term using the particle number density had been proposed for PIC to keep the uniform distribution of the particles (Umegaki et al., 1992). The MPS method inherits this idea.

The particle number density is also useful for the boundary condition of the free surfaces. In the MPS method, the free surfaces are detected by the decrease of the particle number density and the Dirichlet boundary condition $P = 0$ is given to the pressure field when the pressure Poisson equation is solved. This condition is simple and robust even if the fluid is fragmented as well as the free surfaces are largely deformed.

In the MPS method, the principal formulation of spatial discretization is based on the weighted difference which differs from superposition of kernels in the SPH method. Particularly, discretization of the second-order differentiation is simple without using the differentiation of the kernel. Thus, the viscosity term and the Laplacian operator in the pressure Poisson equation can be discretized without complicated processes. The MPS method has been applied to incompressible viscous flows in various engineering problems: civil engineering, coastal engineering, marine engineering, mechanical engineering, automotive engineering, metallurgy, chemical engineering, polymer engineering, food manufacturing, nuclear engineering, etc.

A semi-implicit algorithm was also employed with a pressure Poisson equation in SPH (Cummins and Rudman, 1999). The second-order differentiation was discretized to the combination of the first-order differentiation of the kernel and the first-order difference between two neighboring particles. The source term was given by the divergence of the velocity as the same way as the finite volume method. The weight density, which was almost equivalent to the particle number density, was also used in the source term of the pressure Poisson equation in SPH (Lo and Shao, 2002; Shao and Lo, 2003). The SPH method using the semi-implicit algorithm with the pressure Poisson equation is called ISPH.

On the other hand, the MPS method was also used with a weakly compressible approach with a fully explicit algorithm (Shakibaeinia and Jin, 2010). As the same way as WCSPH, the sound speed was artificially reduced to enhance the numerical stability and the time step was kept large as that of the semi-implicit algorithm.

Now, as explained in Section 1.2.4, the difference between MPS and SPH is the formulation of the spatial discretization. Both the incompressible semi-implicit algorithm and the weakly compressible explicit algorithm are used in both MPS and SPH. For various practical problems in fluid dynamics, various additional techniques are necessary: e.g., the surface tension models, phase change models, and non-Newtonian viscosity calculation. They may be common in MPS and SPH.

The textbooks describing comprehensive knowledge of the particle methods for continuum mechanics are Liu and Liu (2003), Li and Liu (2004), Gotoh (2004), Koshizuka (2005), Koshizuka et al. (2008), Violeau (2012), Gotoh et al. (2013), Koshizuka et al. (2014), Liu and Liu (2016), Yagawa and Sakai (2016), and Gotoh (2017).

REFERENCES

Cummins, S.J., Rudman, M., 1999. An SPH projection method. J. Comput. Phys. 152, 584−607.

Daly, B.J., Harlow, F.H., Welch, J.E., Wilson, E.N., Sanmann, E.E., 1965. Numerical fluid dynamics using the particle-and-force method; Part I: The method and its applications, Part II: Some basic properties of particle dynamics, LA-3144.

Gingold, R.A., Monaghan, J.J., 1977. Smoothed particle hydrodynamics: theory and application to non-spherical stars. Mon. Not. R. Astr. Soc. 181, 375−389.

Gotoh, H., 2004. Computational Mechanics of Sediment Transport. Morikita Shuppan, Tokyo, Japan (in Japanese).

Gotoh, H., Okayasu, A., Watanabe, Y., 2013. Computational Wave Dynamics. World Scientific Publishing.

Gotoh H., 2017. Ryushiho. Morikita Shuppan, (in Japanese).

Harlow, F.H., 1963. Theory of correspondence between fluid dynamics and particle-and-force models, LA-2806.

Harlow, F.H., 1988. PIC and its progeny. Comput. Phys. Commun. 48, 1−10.

Harlow, F.H., Meixner, B.D., 1961. The particle-and-force computing method for fluid dynamics, LAMS-2567.

Koshizuka, S., 2005. Ryushiho. Maruzen Shuppan (in Japanese).

Koshizuka, S., Oka, Y., 1996. Moving-particle semi-implicit method for fragmentation of incompressible fluid. Nucl. Sci. Eng. 123, 421−434.

Koshizuka, S., Harada, T., Tanaka, M., Kondo, M., 2008. Ryushiho Simulation. Baifukan (in Japanese).

Koshizuka, S., Shibata, K., Murotani, K., 2014. Ryushihonyumon. Maruzen Shuppan (in Japanese).

Li, S., Liu, W.K., 2004. Meshfree Particle Method. Springer, Berlin/New York.

Liu, G.R., Liu, M.B., 2003. Smoother Particle Hydrodynamics, a Meshfree Particle Method. World Scientific Publishing.

Liu, M.B., Liu, G.R., 2016. Particle Methods for Multi-Scale and Multi-Physics. Imperial College Press, London.

Lo, E.Y.M., Shao, S., 2002. Simulation of near-shore solitary wave mechanics by an incompressible SPH method. Appl. Ocean Res. 24, 275−286.

Lucy, L.B., 1977. A numerical approach to the testing of the fission hypothesis. Astronom. J. 82, 1013−1024.

Monaghan, J.J., 1988. An introduction to SPH. Comput. Phys. Commun. 48, 89−96.

Monaghan, J.J., 1994. Simulating free surface flows with SPH. J. Comput. Phys. 110, 399−406.

Shakibaeinia, A., Jin, Y.-C., 2010. A weakly compressible MPS method for modeling of open-boundary free-surface flow. Int. J. Numer. Meth. Fluids 63, 1208−1232.

Shao, S., Lo, E.Y.M., 2003. Incompressible SPH method for simulating Newtonian and non-Newtonian flows with a free surface. Adv. Water Res. 26, 787−800.

Umegaki, K., Takahashi, S., Miki, K., 1992. Numerical simulation of incompressible viscous flow using particle method. Suuchikaiseki 43-3, 17−24 (in Japanese).

Violeau, D., 2012. Fluid Mechanics and the SPH Method. Oxford University Press.

Yagawa, G., Sakai, Y., 2016. Ryushiho, Kiso to Oyo. Iwanami Shoten (in Japanese).

CHAPTER 2

Fundamentals of Fluid Simulation by the MPS Method

Abstract

This chapter explains the fundamentals of fluid simulations using the moving particle semi-implicit (MPS) method. Section 2.1 explains the concept of particles using figures and C language programs. Section 2.2 theoretically explains the simulation procedure of the MPS method. Section 2.3 explains sample programs of the MPS method. In Section 2.4, we exercise MPS simulations. If you would like to first experience MPS simulations, it would be better to read Section 2.3 and 2.4 before reading Section 2.2. Section 2.5 lists frequently asked questions and provides answers.

Keywords: Theory of the MPS method; governing equations; discretization; semi-implicit method; source code of sample program; exercise of MPS simulation; FAQ

Contents

2.1	The Elements of the MPS Method	26
	2.1.1 Setting the Initial Positions of Particles	28
	2.1.2 Setting Initial Velocities of Particles	29
	2.1.3 How to Move Particles	29
	2.1.4 How to Calculate Acceleration of Particles	31
2.2	Basic Theory of the MPS Method	33
	2.2.1 Mass of a Particle	33
	2.2.2 Governing Equations	35
	2.2.3 Particle Number Density and Weight Function	40
	2.2.4 Approximation of Partial Differential Operators	45
	2.2.5 Semi-implicit Method	57
2.3	Outline of Simulation Programs	74
	2.3.1 Contents of Program	74
	2.3.2 How to Compile and Execute the Sample Programs	84
	2.3.3 How to Visualize the Simulation Result	85
	2.3.4 Functions of the Program	86
2.4	Exercise of Simulation	103
	2.4.1 Exercises	103
2.5	Hints for Exercises	104
2.6	Frequently Asked Questions	105
	2.6.1 What Is the Best Effective Radius of the Interaction Zone?	105
	2.6.2 Why Do We Need to Arrange Dummy Wall Particles Behind Wall Particles?	105
	2.6.3 Particles Penetrated a Wall: What Is the Possible Reason?	106

Moving Particle Semi-implicit Method
DOI: https://doi.org/10.1016/B978-0-12-812779-7.00002-3

2.6.4 How Do We Set the Time Increment Δt?   106
2.6.5 It Seems a Simulation Diverged Because Particles Exploded: What Is
    the Reason?   107
2.6.6 Fluid Was Compressed: What Is the Reason?   107
2.6.7 How Can We Add a Function for Inlet or Outlet Boundaries in a
    Simulation Program?   107
2.6.8 What Is the Most Time-Consuming Part in an MPS Simulation?   108
  2.6.9 What Are the Drawbacks and the Strong Points of the Semi-implicit Method? 108
References   108

2.1 THE ELEMENTS OF THE MPS METHOD

Fig. 2.1 shows an example of simulation results by the MPS method. We find that a water column, which was stationary at the initial state, starts to break because of gravity. To carry out this type of simulation, first we arrange fluid particles in locations where we would like to locate a fluid region. As shown in Fig. 2.2A, we usually arrange particles at regular intervals. The interval is called the initial distance between particles, and is expressed as l_0.

In Figs. 2.1 and 2.2A, particles are drawn as circles, which have clear surfaces. However, particles in the MPS method are actually calculation points as shown in Fig. 2.2B, and do not have clear surfaces. In the MPS method, we simulate fluid using a group of calculation points. As we

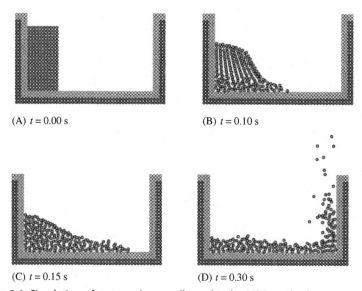

(A) $t = 0.00$ s   (B) $t = 0.10$ s

(C) $t = 0.15$ s   (D) $t = 0.30$ s

Figure 2.1 Simulation of water column collapse by the MPS method.

explained in Chapter 1, Introduction, each calculation point has variables of position, velocity, pressure, etc. We update velocities of particles on the basis of the equation of motion, which expresses fluid motion, and move particles at the updated velocities to express fluid motion.

Note 2.1: *Particle ID*: In Fig. 2.2, the subscript of each particle expresses the particle's identification number (ID), which is called a "particle ID."

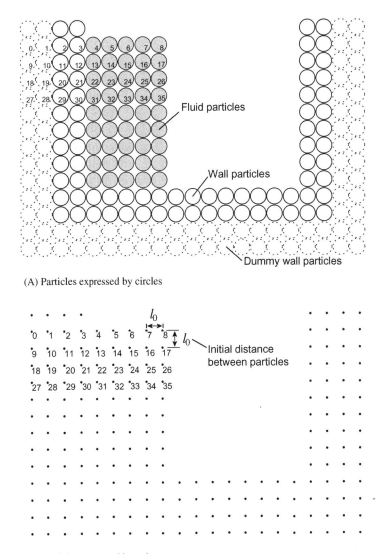

(A) Particles expressed by circles

(B) Particles expressed by points

Figure 2.2 Particle arrangement and expressions: (A) particles expressed by circles and (B) particles expressed by points. *The numbers are the IDs of particles.*

In this book, we call a particle, whose particle ID is i, the "ith particle" or "particle i." For example, the particle whose particle ID is 5 is called 5th particle. The particle IDs start from 0 in the sample programs of this book because the programs are written in the C language, whose subscripts of arrays start from 0.

Note 2.2: *Spatial resolution*: The initial distance between particles, l_0, expresses the spatial resolution of the MPS method. Shorter l_0 expresses higher spatial resolution. The higher spatial resolution requires more particles and longer computation time.

2.1.1 Setting the Initial Positions of Particles

In order to carry out MPS simulations, we first need to determine the positions of particles at the initial state. At that time, we need to consider the spatial resolution and arrange particles regularly. For example, in the case where the 0th, 1st, and 2nd particles are adjacent and the distance between initial particles, l_0, is 0.1 m, we can determine the 0th, 1st, and 2nd particle positions as (0, 0, 0), (0.1, 0, 0), and (0.2, 0, 0), respectively. We can write a C language program of the above procedure as shown in Program 2.1 (Note 2.3).

```
1   PositionX[0] = 0.0;  PositionY[0] = 0.0;  PositionZ[0] = 0.0;
2   PositionX[1] = 0.1;  PositionY[1] = 0.0;  PositionZ[1] = 0.0;
3   PositionX[2] = 0.2;  PositionY[2] = 0.0;  PositionZ[2] = 0.0;
        ...
```

Program 2.1

The arrays, PositionX[ ], PositionY[ ], and PositionZ[ ] are one-dimensional arrays of double type. Elements, PositionX[0], PositionY [0], and PositionZ[0] express coordinates of the 0th particle in x, y, and z directions, respectively. We set the coordinates of all the particles using a loop statement, such as the "for" statement of the C language to shorten the program, although a loop statement is not used in Program 2.1 to simplify the explanation (Note 2.4).

Note 2.3: *Declaration of variables*: It is necessary to declare types of arrays or variables before using them in a program. For example, if you use 5000 particles, a statement "double PositionX[5000], PositionY[5000], PositionZ[5000];" is necessary before executing Program 2.1. In this book, we explain programs omitting this type of declaration.

Note 2.4: *"Double" type*: "Double" is a type of value and is the abbreviation for "double precision floating-point values."

2.1.2 Setting Initial Velocities of Particles

It is also necessary to set up the initial velocities of particles. Program 2.2 is an example C language program for setting the initial velocities.

```
1   VelocityX[0] = 0.0;  VelocityY[0] = 0.0;  VelocityZ[0] = 0.0;
2   VelocityX[1] = 0.0;  VelocityY[1] = 0.0;  VelocityZ[1] = 0.0;
3   VelocityX[2] = 0.0;  VelocityY[2] = 0.0;  VelocityZ[2] = 0.0;
                              ...
```

Program 2.2

Arrays VelocityX[ ], VelocityY[ ], and VelocityZ[ ] are one-dimensional arrays of double type. Elements, VelocityX[0], VelocityY[0], and VelocityZ[0] express the component of velocity of the 0th particle in x, y, and z directions, respectively. In this case, all the velocity components of the 0th, 1st, and 2nd particles are 0.0 m/s. That expresses these particles are stationary at the initial state. We carry out similar procedures for all the particles to set their initial velocities (Note 2.5).

Computers are good at expressing a continuous value as a group of small-segment values. For example, in the MPS method, a liquid fluid is divided into small particles. A particle is a type of segment. As mentioned in Section 1.1.1, each particle has information about position, velocity, pressure, etc. That is, fluid region, velocity distribution, and pressure distribution which are continuous quantities are divided into segments. In most simulation programs, a group of segment values is expressed by an array as shown in Programs 2.1 and 2.2.

Note 2.5: *Initial velocity of fluid*: We can set any value for the initial velocity of fluid as well as 0.0 m/s.

2.1.3 How to Move Particles

Before moving particles, first we update the velocities of particles because particle velocities might be accelerated or decelerated after the previous time step. Then, we calculate displacements of particles by multiplying the updated particle velocity and the time increment Δt. The direction of the particle displacement is the same as that of the particle velocity.

Next, we explain the above-mentioned procedures by a C language program. We can write a C program that calculates the velocity of ith particle at the next time step as shown in Program 2.3.

```
1  DT = 0.001;
2  VelocityX[i] = VelocityX[i] + AccelerationX[i] * DT;
3  VelocityY[i] = VelocityY[i] + AccelerationY[i] * DT;
4  VelocityZ[i] = VelocityZ[i] + AccelerationZ[i] * DT;
```

Program 2.3

AccelerationX[i], AccelerationY[i], and AccelerationZ[i] are the components of the acceleration vector of the ith particle in x, y, and z directions, respectively. The variable DT is the time increment Δt, and is set to 0.001 s in the first line of Program 2.3. In a simulation program, space and time are expressed as discrete values as described in Chapter 1, Introduction. In the above program, " $=$ " expresses substituting the right-hand side for the left-hand side. The asterisk, *, expresses multiplication. In the right-hand side of the second line, the component of the velocity increment in the x-direction (AccelerationX[i] * DT) is added to the x-directional component of the present velocity vector, which is expressed by VelocityX[i] on the right-hand side in line 2 of Program 2.3. The calculation result is stored in VelocityX[i] on the left-hand side in line 2. By this procedure, we can update the velocity component in x-direction. In the same manner, the velocity components in y and z directions are updated in lines 3—4, and the particle velocity of the next time step is obtained. Although Program 2.3 is written for only the ith particle, we actually carry out the same procedure for all the particles.

Finally, we can move particles by adding the displacement of each particle to the current particle position to obtain the particle positions of the next time step as shown in Program 2.4.

```
1  PositionX[i] = PositionX[i] + VelocityX[i] * DT;
2  PositionY[i] = PositionY[i] + VelocityY[i] * DT;
3  PositionZ[i] = PositionZ[i] + VelocityZ[i] * DT;
4
```

Program 2.4

Although Program 2.4 is written for only the ith particle, we actually carry out the same procedure for all the fluid particles and move the particles.

2.1.4 How to Calculate Acceleration of Particles

As mentioned earlier, we can calculate the displacement of fluid particles by simple calculations as shown in Programs 2.3 and 2.4 provided we can obtain the acceleration of each particle. We can obtain the particle accelerations by the equation of motion, which is expressed as follows:

$$m\mathbf{a} = \mathbf{F} \tag{2.1}$$

Here, m is the mass of an object, $\mathbf{a}$ is the translational acceleration of the object, and $\mathbf{F}$ is the force acting on the object. By dividing both sides by m, Eq. (2.1) is expressed as follows:

$$\mathbf{a} = \frac{\mathbf{F}}{m} \tag{2.2}$$

From the equation, the equation of motion expresses the acceleration of a particle, and we can calculate the acceleration vector by dividing the force by the mass of the object. Moreover, note that the acceleration is in proportion to $\mathbf{F}$, and is in inverse proportion to m.

The equations of motion for fluid are the Navier—Stokes equations as explained in Section 1.1.3. Let us review the incompressible Navier—Stokes equations, and confirm that it is expressed in the same form as Eq. (2.1). The incompressible Navier—Stokes equation is expressed as follows:

$$\rho \frac{D\mathbf{u}}{Dt} = -\nabla P + \mu \nabla^2 \mathbf{u} + \rho \mathbf{g} \tag{2.3}$$

where ρ is the fluid density, $\mathbf{u}$ is the fluid velocity, $\frac{D}{Dt}$ is the material derivative, ∇ is the nabla, P is pressure, μ is the viscosity coefficient, ∇^2 is the Laplacian, and $\mathbf{g}$ is the acceleration of gravity. The right-hand side of Eq. (2.3) is composed of forces acting on a fluid which has unit volume. The first, second, and third terms on the right-hand side are the pressure gradient term, the viscous term, and the gravity term, respectively. The details of each term are explained in Section 2.2.2.1. Although the Navier—Stokes equation appears difficult to understand, the equation becomes easy to understand by multiplying both sides by a constant infinitesimal volume V as follows (Note 2.6):

$$\rho V \frac{D\mathbf{u}}{Dt} = \left(-\nabla P + \mu \nabla^2 \mathbf{u} + \rho \mathbf{g}\right) V \tag{2.4}$$

The coefficient ρV on the left-hand side of Eq. (2.4) expresses the mass of a particle whose volume is infinitesimal. The material derivative of velocity $\frac{D\mathbf{u}}{Dt}$ is equal to the acceleration vector of a fluid particle. The material derivative is called the Lagrange derivative or substantial derivative, and is explained in Section 2.2.5.3 and Notes 2.15 and 2.16 in detail. The material derivative expresses the time change of a physical quantity of a material point, where a material point is a fluid particle in the MPS method. That is, $\mathbf{a} = \frac{D\mathbf{u}}{Dt} = \frac{D^2\mathbf{r}}{Dt^2}$, where the vector $\mathbf{r}$ is the position of a particle and is expressed as $\mathbf{r} = \left(r_x, r_y, r_z\right)^T$. The scalar parameters r_x, r_y, and r_z are the coordinate components of a particle in x, y, and z directions, respectively. The right-hand side of Eq. (2.4) expresses the force acting on a particle because it is the multiplication of $\left(-\nabla P + \mu\nabla^2\mathbf{u} + \rho\mathbf{g}\right)$, which is the force acting on a unit volume, by the volume of particle V. Thus, the right-hand side of Eq. (2.4) indicates the resultant force vector $\mathbf{F}$ acting on a particle. Therefore, we can find that the right-hand side of Eq. (2.4) is $\mathbf{F}$, and the left-hand side is $m\mathbf{a}$. This relationship is the same as Eq. (2.1) (Note 2.7).

If both sides of Eq. (2.4) are divided by ρV, which is the mass of a fluid particle, we obtain an equation similar to Eq. (2.2) as follows. (Important equations, which are written in a simulation program, are boxed in this chapter to emphasize their importance.)

$$\frac{D\mathbf{u}}{Dt} = -\frac{1}{\rho}\nabla P + \nu\nabla^2\mathbf{u} + \mathbf{g} \tag{2.5}$$

where ν is the kinematic coefficient of viscosity and is defined by $\nu = \mu/\rho$. The left-hand side of Eq. (2.5) expresses the acceleration of a fluid particle, and the right-hand side expresses the force vector acting on a fluid particle, which is divided by the mass of the fluid particle. By calculating the right-hand side of Eq. (2.5), the acceleration vector of a fluid particle, which is the left-hand side, is obtained. Using the calculated accelerations of particles, we can update particle velocities and move particles at the updated velocities by the easy processing shown in Programs 2.3 and 2.4.

There are two important notes. First, the pressure in Eq. (2.5) is an unknown value, and the acceleration of fluid particles cannot be calculated only by Eq. (2.5). This problem is solved by introducing "the equation of continuity" or "the equation of state" as another conditional equation. The details are explained in Section 2.2.5.1 and Note 2.18. Another note is that the partial differential operators, such as nabla ∇ and Laplacian ∇^2, in Eq. (2.5) need to be replaced with easier operators. This is because computers do not directly differentiate or integrate functions. Therefore, we need to substitute these differential operators with basic operators such as the four basic operations of arithmetic (addition, subtraction, multiplication, and division), which are easily calculated by computers. Usually this procedure of substitution divides continuous quantities into small segments of space or time, and is called "discretization." Each segment has a data set which expresses the quantities of the segment. The accuracy, robustness, simplicity, and cost of simulations depend on the discretization method. Many researches have been working on improving discretization methods. The MPS method is one such method, and we replace the operators of partial differentiation with particle models. The details are explained in Section 2.2.4.

Note 2.6: *Volume integration*: The multiplication of both sides of Eq. (2.4) by the volume V of a particle expresses an approximation of volume integration. This approximation is valid if the particle volume is very small.

Note 2.7: *Transposition symbol T*: The superior letter, T, expresses the transposition of a vector or matrix. We express column vectors using transposition to save space. For example, we express a position vector $\mathbf{r}$ as

$$\mathbf{r} = \begin{pmatrix} r_x \\ r_y \\ r_z \end{pmatrix} = \left(r_x, r_y, r_z \right)^{\mathrm{T}}.$$

2.2 BASIC THEORY OF THE MPS METHOD

2.2.1 Mass of a Particle

As mentioned earlier, particles in the MPS method are calculation points. Particles do not actually have a clear surface because we can carry out simulations without defining the clear surface of each particle. If all particles are the same in size, we can simulate a fluid with particles without considering the mass of each particle because mass is indirectly considered by the fluid density ρ. However, we explain the mass of a particle below because it will help you to understand the theory of the MPS method in detail.

For example, we consider a case where the fluid is water, and the total volume of fluid is 1 m³. The fluid density of water is 1000 kg/m³, and the total fluid mass of the fluid is thus 1000 kg. If the fluid is discretized by 1000 fluid particles in the MPS method, the volume per one particle is 1×10^{-3} m³, and the mass of a particle is 1 kg. In order to express the mass of a particle by the above-mentioned initial distance between particles l_0, we first express the volume of a particle in terms of l_0 because we can calculate the mass of a particle by multiplying the fluid density and the particle volume. In the case of a three-dimensional simulation where particles are arranged at regular intervals as shown in Fig. 2.2, each particle has a volume of l_0^3 m³ because each particle represents a small cubic domain whose sides are l_0 m. We can obtain the mass of a particle by multiplying fluid density and the volume per particle. Therefore, the mass of a particle is ρl_0^3 kg in the case of a three-dimensional simulation. In the same manner, the mass of a particle in two dimensions is ρl_0^2 kg/m because we consider the mass of a particle per unit depth. In a two-dimensional simulation, we assume that shape and physical quantities of fluid are the same in the depth direction, and we omit calculations about the depth direction by solving for fluid per unit depth. In the MPS method, the mass of a particle is treated as a constant parameter, which does not change over time.

Note 2.8: *Conservation of mass in the MPS method*: In the MPS method, the conservation of mass is satisfied perfectly because the mass of a particle and the number of particles do not change as time passes, unless we delete or generate particles. Although it seems natural in the case of particle methods, it needs careful attention in grid-based methods to perfectly satisfy mass conservation. The intrinsic feature of the conservation of mass is one of the major advantages of the MPS method.

Note 2.9: *Particle shape*: We calculated the equivalent volume of a particle as if the particle shape is cubic in three dimensions. However, the actual shape is not cubic. The actual shape of a particle is expressed by a weight function, which is explained later. As the surface shape is not defined clearly, we consider each particle as a sphere in three-dimensional simulations or a circle in two-dimensional simulations. Although the surface of each particle is not clearly defined, we can determine the water level of a simulation result by finding the particle located at the highest position and adding $0.5l_0$ to the highest position, where $0.5l_0$ is the half of the initial distance between particles. We can omit the $0.5l_0$ shift by arranging all particles upward by $0.5l_0$ in advance because water level is a relative length from the bottom. Because the sample programs in this chapter also use $0.5l_0$ shifting, we do not need the height calibration.

2.2.2 Governing Equations

In the MPS method, we move particles on the basis of fundamental equations called the governing equations which determine the rules of phenomena in the same manner as in other simulation methods. The governing equations of the MPS method are the Navier–Stokes equations (conservation of momentum) and the equation of continuity (conservation of mass) which are the governing equations of a fluid. Details are given as follows.

2.2.2.1 The Navier–Stokes Equations

Let us take a look at the incompressible Navier–Stokes equation again. The equation is expressed as follows:

$$\frac{D\mathbf{u}}{Dt} = -\frac{1}{\rho}\nabla P + \nu\nabla^2\mathbf{u} + \mathbf{g} \tag{2.6}$$

where $\frac{D\mathbf{u}}{Dt}$ is the material derivative of fluid velocity and expresses the acceleration of a fluid particle. Eq. (2.6) indicates that the fluid acceleration consists of the acceleration components on the right-hand side. These acceleration components are explained as follows.

2.2.2.1.1 Meaning of the Pressure Gradient Term

The first term on the right-hand side of Eq. (2.6), $-\frac{1}{\rho}\nabla P$, is the acceleration due to pressure. As this term is in proportion to ∇P, which expresses a gradient (slope) of pressure field, we find that this term is an acceleration in proportion to the gradient of the pressure field. Therefore, this term is called a pressure gradient term or a pressure term. For the details of the operator ∇, see Section 2.2.4. The pressure gradient is divided by the fluid density ρ, which is the mass per unit volume. The division by ρ is necessary to express the effect of fluid mass per unit volume. If we consider the case where fluid densities are different under the same pressure gradient, the pressure gradient term of the lower density fluid is higher than that of the higher density fluid. There is a negative sign " $-$ " in the term $-\frac{1}{\rho}\nabla P$. This is for turning the acceleration in the direction where the pressure is the lowest around the position of a certain fluid particle. Because the gradient of pressure ∇P points in the direction where the pressure is the highest around the position of the fluid particle, we need to reverse the direction. That is why the negative sign is necessary for $-\frac{1}{\rho}\nabla P$. We can derive the negative sign by considering the direction of the fluid movement and the difference between pressures in each direction.

We can compare this pressure gradient term to a displacement of a person in a crowded train. Let us suppose you are in a crowded train. Your right-hand side is very crowded. This situation indicates that your right-hand side has a very high pressure. On the other hand, your left-hand side is not so crowed. This is similar to that your left-hand side experiencing a low pressure. You might collide with other people, and be pushed from both sides. Because your right-hand side is more crowded than the left-hand side, your body will be moved leftward by the collisions. This situation is similar to the fluid simulation, where a fluid particle is accelerated in the direction from the high pressure place to the low pressure place. Of course, in the actual fluid phenomena, collisions occur not between people but between fluid particles.

2.2.2.1.2 Meaning of the Viscous Term

The second term on the right-hand side of Eq. (2.6) is the viscous term, which is the acceleration caused by the viscous force. The viscous term is a term which diffuses the momentum of fluid because the velocity components will change in proportion to the result of the calculation of the Laplacian in common with the quantity expressed by the diffusion equation. For the details of the Laplacian operator, refer to Section 2.2.4.6. In other words, the viscous term expresses the force which is going to make the fluid velocity of a particle similar to the velocities of neighboring fluid particles. Fluid consists of many molecules. The molecules interact with each other by collisions and intermolecular forces. These interactions generate frication effects. Consider a certain fluid particle i among particles. To simplify the problem, consider the case where all the particles are moving in the x-direction. If the velocities of the neighboring particles around the ith particle are faster than the ith particle, the particle i will be dragged by surrounding particles according to the frictional force. As a result, particle i will accelerate in the x-direction. In contrast, if surrounding particles are moving at a slower velocity than particle i, the velocity of the ith particle will decelerate in the x-direction because of the friction force. The viscous term expresses such friction effects. The friction works between a fluid particle and a wall as well as between fluid particles. The viscous term has the effect of attenuating the kinetic energy of fluid (Note 2.10).

Moreover, from the form of $\nu\nabla^2\mathbf{u}$, we find that the acceleration due to the viscous force is in proportion to the second derivative of a velocity because ∇^2 is expressed as $\frac{\partial^2}{\partial x^2} + \frac{\partial^2}{\partial y^2} + \frac{\partial^2}{\partial z^2}$. A second partial differentiation of a physical quantity with respect to the x, y, or z coordinate expresses a type of curvature of the distribution of a quantity in each direction. For example, let us consider the case where a physical quantity is approximated as a function of $f \cong ax^2 + bx + c$, where a, b, and c are the coefficients of a polynomial. The curvature of the distribution is large if the coefficient "a" is large. We can confirm this by drawing a graph of the function in the simple case, where b and c are zero and f is expressed as $f \cong ax^2$. If the coefficient "a" is large, the curve on the graph has a large curvature. If we calculate the second differential derivative of f about x, $\frac{\partial^2}{\partial x^2} f \cong a$. This result agrees with the earlier discussion that the second derivative of a function expresses a type of a curvature. The coefficients b and c do not affect the calculation result of second derivatives because the second derivative of $f \cong bx + c$ is approximately zero. Even if the velocity is very high, if the velocity distribution is expressed as $f \cong bx + c$, the viscosity is very small because the second differential derivative is $\frac{\partial^2}{\partial x^2} f \cong 0$. Even if the velocity is very low, the viscous term can be large for a high curvature of the velocity distribution or a large kinematic viscosity ν. The kinematic viscosity coefficient has been studied for various fluids. The coefficient depends on the type of fluid and its temperature. The kinematic viscosity of water is about $1.00 \times 10^{-6} \mathrm{m}^2/\mathrm{s}$ at $20°C$. For the kinematic viscosity coefficients of other fluids, see reference Kaye & Laby. Velocity distributions often have large curvature on wall boundaries unless the walls are slippery. As a result, the viscosity force tries to reduce the curvature of the velocity distribution (Note 2.11).

Note 2.10: *Diffusion equation*: A diffusion equation expresses a diffusion process, and is expressed as $\frac{D\Phi}{Dt} = C\nabla^2\Phi$, where Φ is an arbitrary function and C is a diffusion coefficient.

Note 2.11: *Velocity on a wall*: On a nonslip wall boundary, where fluid does not slip along the wall, the velocity on the wall surface is treated as 0 m/s.

2.2.2.2 Equation of Continuity

The equation of continuity is expressed as follows:

$$\frac{D\rho}{Dt} + \rho\nabla\cdot\mathbf{u} = 0 \tag{2.7}$$

The equation of continuity expresses the law of mass conservation. Specifically, the equation of continuity expresses how fluid density changes according to the mass flow from a certain unit volume. The first term on the left-hand side of Eq. (2.7) expresses the time change of fluid density per unit domain. The second term on the left-hand side is the multiplication of the fluid density by the divergence of velocity. The divergence of velocity expresses the fluid volume which flows out from a unit volume per unit time. Therefore, the second term expresses the fluid mass that flows out from the unit volume per unit time because the fluid density is multiplied by the volume. (Although it appears necessary to divide the mass by the volume to calculate the density, the division is not necessary because the equation is considered per unit volume, and has already been divided by the volume.) We can make Eq. (2.7) easy to understand by transforming it as follows:

$$\frac{D\rho}{Dt} = -\rho\nabla\cdot\mathbf{u} \tag{2.8}$$

From the above equation, we can easily understand that the fluid density decreases if the fluid mass flows out from the unit volume because $\nabla\cdot\mathbf{u}$ is positive, and the material derivative of fluid density becomes negative because coefficient $-\rho$ is multiplied by $\nabla\cdot\mathbf{u}$. The negative sign is necessary for the right-hand side because the fluid density increases in the case where fluid flows in, while it decreases when fluid flows out. If Eq. (2.6) is not satisfied completely, some part of the fluid mass is lost, or the fluid mass is increased. Therefore, it is very important to satisfy the equation of continuity in grid-based methods. On the other hand, the MPS method can satisfy mass conservation very easily because the MPS method assumes every particle has a constant mass. If we do not generate or delete particles, the local and total fluid masses are completely conserved. This is an advantage of the MPS method. In the MPS method, the equation of continuity is indirectly considered in the calculation of pressure Poisson equation. The details are explained in Section 2.2.5. For the more detailed contents of hydrodynamics, see the references Hughes and Brighton (1999) and Potter and Wiggert (2007).

2.2.2.3 Notation by Vectors

The governing equations (Eqs. 2.6 and 2.7) are expressed in vector notation. Actually, these equations have x, y, and z directional components. For those who are not familiar with vector notation yet, we would like to explain the equations using x, y, and z directional components. First, Eq. (2.6) is expressed by components as follows:

$$\rho \frac{Du_x}{Dt} = -\frac{\partial P}{\partial x} + \mu \left(\frac{\partial^2 u_x}{\partial x^2} + \frac{\partial^2 u_x}{\partial y^2} + \frac{\partial^2 u_x}{\partial z^2} \right) \tag{2.9}$$

$$\rho \frac{Du_y}{Dt} = -\frac{\partial P}{\partial y} + \mu \left(\frac{\partial^2 u_y}{\partial x^2} + \frac{\partial^2 u_y}{\partial y^2} + \frac{\partial^2 u_y}{\partial z^2} \right) - \rho g \tag{2.10}$$

$$\rho \frac{Du_z}{Dt} = -\frac{\partial P}{\partial z} + \mu \left(\frac{\partial^2 u_z}{\partial x^2} + \frac{\partial^2 u_z}{\partial y^2} + \frac{\partial^2 u_z}{\partial z^2} \right) \tag{2.11}$$

where u_x, u_y, and u_z are the velocity components in the x, y, and z directions. In the derivation of the above equations, we used the following definitions:

$$\mathbf{u} = \left(u_x, u_y, u_z \right)^{\text{T}} \tag{2.12}$$

$$\nabla \equiv \begin{pmatrix} \dfrac{\partial}{\partial x} \\[2mm] \dfrac{\partial}{\partial y} \\[2mm] \dfrac{\partial}{\partial z} \end{pmatrix} \tag{2.13}$$

$$\nabla^2 \equiv \nabla \cdot \nabla = \begin{pmatrix} \dfrac{\partial}{\partial x} \\[2mm] \dfrac{\partial}{\partial y} \\[2mm] \dfrac{\partial}{\partial z} \end{pmatrix} \cdot \begin{pmatrix} \dfrac{\partial}{\partial x} \\[2mm] \dfrac{\partial}{\partial y} \\[2mm] \dfrac{\partial}{\partial z} \end{pmatrix} = \dfrac{\partial^2}{\partial x^2} + \dfrac{\partial^2}{\partial y^2} + \dfrac{\partial^2}{\partial z^2} \tag{2.14}$$

$$\rho\mathbf{g} \equiv (0, -\rho g, 0)^{\mathrm{T}} \tag{2.15}$$

where g is the acceleration of gravity and is set to 9.81 m/s. There is a negative sign in the y component of $\rho\mathbf{g}$ because the positive direction of the y coordinate is the upward vertical direction in this chapter.

In the same manner, Eq. (2.6) is expressed by components as follows:

$$\frac{D\rho}{Dt} + \rho\left(\frac{\partial u_x}{\partial x} + \frac{\partial u_y}{\partial y} + \frac{\partial u_z}{\partial z}\right) = 0 \tag{2.16}$$

Compared to the component expressions of Eqs. (2.9), (2.10), (2.11), and (2.16), the vector expressions of Eqs. (2.6) and (2.7) are concise and are often used. This book also often expresses equations by vectors. Note that there are equations and terms for all the components.

2.2.3 Particle Number Density and Weight Function

Fluid density is important for fluid simulations because mass conservation is expressed by the fluid density (Eq. 2.7), and the pressure is calculated by fluid density. In gas simulations, where fluids are easy to compress, pressure is often calculated by the equation of state, which is expressed as a function of fluid density. The details are explained in Section 2.3.4.17, Program 2.21, and Section 3.2 (Eq. 3.38). In liquid simulations, fluids can be treated as incompressible in many cases. We often add a condition that fluid does not compress and fluid density is constant. In this case, the relationship for fluid density (Eq. 2.7) is used.

Thus, it is necessary to appropriately evaluate fluid density to express fluid phenomena. In the MPS method, we use the particle number density to evaluate the fluid density. The particle number density reflects the number of particles around a particle. We define the particle number density as the summation of neighboring particles' weight. The particle

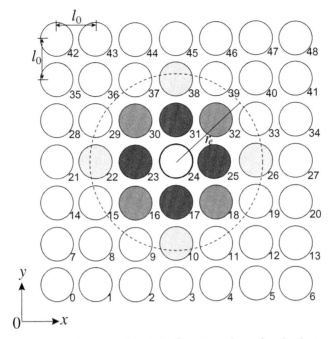

Figure 2.3 Schematic diagram of weight function. *The color depth expresses the weight of each particle. The numbers are IDs of particles.*

number density expresses the density of particles at a particle position. Fig. 2.3 shows a conceptual image of the calculation of the particle number density at the position of a certain particle i. The color depth in the figure expresses the importance of the significance. The deep color expresses a large weight, and the light color expresses a light weight. As shown in the figure, particles which are near particle i have larger weight.

Specifically, we calculate the particle number density of the ith particle by the following equation (Koshizuka and Oka, 1996):

$$n_i = \sum_{j \neq i} w\left(\left|\mathbf{r}_j - \mathbf{r}_i\right|, r_e\right) \tag{2.17}$$

where n_i is the particle number density of the ith particle, $\mathbf{r}_i$ is the position vector of particle i, and $\mathbf{r}_j$ is the position vectors of neighboring particles j. The function $w(\ )$ is the weight function and is explained later. The symbol $\sum_{j \neq i}$ expresses the summation of all the j particles other than i. If the distance between particles exceeds the effective radius r_e, the weight becomes zero. To write an efficient program, we should calculate the

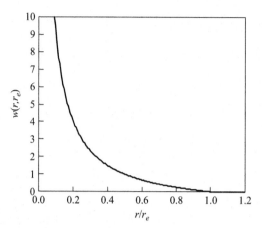

Figure 2.4 Weight function.

summation about only the neighboring particles because the weight function becomes zero if the distance between particles is longer than or equal to the effective radius. The symbol ‖ expresses the absolute value, and $|\mathbf{r}_j - \mathbf{r}_i|$ indicates the distance between particle i and particle j. The weight function $w(r, r_e)$ is calculated as follows:

$$w(r, r_e) = \begin{cases} \left(\dfrac{r_e}{r}\right) - 1 & (r < r_e) \\ \\ 0 & (r \geq r_e) \end{cases} \qquad (2.18)$$

where r expresses the distance between particles, and r_e is the effective radius of the interaction zone. The parenthesis of $w(r, r_e)$ means that the weight function w is a function of r and r_e. Fig. 2.4 expresses the shape of Eq. (2.17). We find that the function decreases as the distance between particles increases.

2.2.3.1 The Standard Particle Number Density n^0

As shown in Fig. 2.3, the standard particle number density is calculated about a certain internal particle i' at the initial state ($t = 0$ s) of a simulation under the condition where particles are arranged at equal intervals. The internal particle i' is a particle which is in the inside of the fluid region (e.g., 24th particle in Fig. 2.3). That is, we cannot choose the 48th particle as the particle i' because it is near a free surface. This is because the particles near a free surface do not have particles on all sides, so the particle number density of the 48th particle is lower than the 24th particle. We express the standard particle number density as n^0 and use it as the standard value of the particle number density. Specifically, we calculate n^0 by the following formula:

$$n^0 = \sum_{j \neq i'} w\left(\left|\mathbf{r}_j^0 - \mathbf{r}_i^0\right|, r_e\right) \tag{2.19}$$

The variables with zero superior express the values at the initial state $(t = 0 \text{ s})$, where particles are arranged at regular intervals. We use the calculated n^0 as a common and constant value among all the particles during a simulation. Moreover, n^0 is also the standard value of the sum total of weights, and is also used as a constant for normalizing the weight function of a weighted average. Note 2.13 in Section 2.2.4.3 describes the details.

2.2.3.2 Relationship Between Particle Number Density and Fluid Density

We can calculate the density change and volume change of fluid using the particle number density. For example, if n_i^k, which is the particle number density of a certain particle i at time step k, is higher than n^0, we find that the fluid density is higher than the standard value, and the fluid is compressed. In contrast, if the particle number density is lower than n^0, we find that the fluid density is lower than the standard value, and the fluid swells. In the MPS method, we calculate the change of fluid density by the following approximation.

$$\frac{\rho_i^k - \rho^0}{\rho^0} \cong \frac{\left(mn_i^k / V_e\right) - \left(mn^0 / V_e\right)}{mn^0 / V_e} = \frac{n_i^k - n^0}{n^0} \tag{2.20}$$

where ρ^0 is the initial fluid density used as the standard value. For example, in the case of liquid water, ρ^0 is about 1000 kg/m^3 at $20°C$ under atmospheric pressure. The variable ρ_i^k is the fluid density of a fluid particle i at the kth time step under a certain pressure. Constant m is the mass of each particle. Constant V_e is the volume of the interaction zone. The constant volume V_e is $\frac{4}{3}\pi r_e^3$ of a sphere in a three-dimensional simulation, where r_e is the effective radius of the interaction zone. In two-dimensional simulations, the effective zone is cylindrical, and V_e is $\pi r_e^2 \times 1$, where a depth of 1 m is multiplied by the area of a circle because we assume the depth is per unit length in two-dimensional simulations. Eq. (2.19) is an approximation, and expresses the degree of compression at the position of the ith fluid particle at the time step k. By Eq. (2.19), we can evaluate the degree of fluid compression by only the particle number density.

2.2.3.3 Example of Calculation

In order to ensure our understanding so far, let us calculate the value of a weight function and a particle number density concretely. We assume that the fluid is two-dimensional, and fluid particles are arranged at regular intervals as shown in Fig. 2.3. The distance between particles l_0 is 0.1 m, and the effective radius of the interaction zone is $2.1l_0$. Now, we calculate n_{24}, which is the particle number density of the 24th particle, as an example. There are particles whose IDs are 10, 16, 17, 18, 22, 23, 25, 26, 30, 31, 32, and 38 in the interaction zone. These particles are expressed by circles. From Eqs. (2.17) and (2.18), we can calculate the particle number density of the 24th particle as follows:

$$
\begin{aligned}
n_{24} &= \sum_{j \neq 24} w\left(\left|\mathbf{r}_j - \mathbf{r}_{24}\right|, r_e\right) \\
&= w\left(\left|\mathbf{r}_{10} - \mathbf{r}_{24}\right|, r_e\right) + w\left(\left|\mathbf{r}_{16} - \mathbf{r}_{24}\right|, r_e\right) + w\left(\left|\mathbf{r}_{17} - \mathbf{r}_{24}\right|, r_e\right) \\
&\quad + w\left(\left|\mathbf{r}_{18} - \mathbf{r}_{24}\right|, r_e\right) + w\left(\left|\mathbf{r}_{22} - \mathbf{r}_{24}\right|, r_e\right) + w\left(\left|\mathbf{r}_{23} - \mathbf{r}_{24}\right|, r_e\right) \\
&\quad + w\left(\left|\mathbf{r}_{25} - \mathbf{r}_{24}\right|, r_e\right) + w\left(\left|\mathbf{r}_{26} - \mathbf{r}_{24}\right|, r_e\right) + w\left(\left|\mathbf{r}_{30} - \mathbf{r}_{24}\right|, r_e\right) \\
&\quad + w\left(\left|\mathbf{r}_{31} - \mathbf{r}_{24}\right|, r_e\right) + w\left(\left|\mathbf{r}_{32} - \mathbf{r}_{24}\right|, r_e\right) + w\left(\left|\mathbf{r}_{38} - \mathbf{r}_{24}\right|, r_e\right) \\
&= 4 \times \left\{ w(0.1, \ 2.1l_0) + w\left(0.1 \times \sqrt{2}, \ 2.1l_0\right) + w(0.2, \ 2.1l_0) \right\} \\
&= 4 \times \left\{ \left(\frac{2.1 \times 0.1}{0.1} - 1 \right) + \left(\frac{2.1 \times 0.1}{0.1 \times \sqrt{2}} - 1 \right) + \left(\frac{2.1 \times 0.1}{0.2} - 1 \right) \right\} \\
&\cong 6.539696962
\end{aligned}
$$

(2.21)

We do not add particle 24 to the set of neighborhood particles. In this example, particles are arranged at equal intervals, and we can use the particle number density of the 24th particle as the standard particle number density n^0.

If you have time to write a simulation program of the MPS method, you may wish to confirm whether the calculated n^0 of your program agrees with the above value. Usually, $1.5l_0$ to $3.1l_0$ is used for the effective radius r_e. Although the value of particle number density depends on the effective radius, we seldom change the effective radius for different problems because the particle number density is used as a form of variation rate as shown in the right-hand side of Eq. (2.20).

Note 2.12: *How to judge the inside and outside of the interaction zone*: As shown in Fig. 2.2B, particles are calculation points and actually do not have clear circular surfaces although particles are usually drawn as circles in Figs. 2.2A and 2.3 to clearly visualize fluid shapes. Therefore, we use the center position of a neighboring particle to judge whether the particle is in the interaction zone.

2.2.3.4 The Form of a Weight Function

Let us consider the form of the weight function shown in Fig. 2.4. We find that the weight decreases as the distance between particles becomes longer. Therefore, we can reduce the influence of distant particles. The form of the weight function has other advantages. As the fluid is expressed by particles, the number of particles in an interaction zone can increase or decrease if some particles flow into or flow out of the interaction zone. As a result, the particle number density fluctuates, and the fluctuation is mistakenly evaluated as compression or expansion of fluid. The weight function of Eq. (2.18) can reduce the fluctuation because the weight becomes low around the area where the distance from the ith particle is about r_e. Moreover, the form of the weight function makes it easy to detect local compression because the weight becomes large if the distance between particles is short. The increase in weight can restrain a fluid particle from passing through a wall or other fluid particles. Eq. (2.18) is usually used as the standard weight function because of its simple form, although other equations are available.

2.2.4 Approximation of Partial Differential Operators

The Navier–Stokes equations, which are the fluid equations of motion for fluid, contain partial differential operators: the nabla ∇ and Laplacian ∇^2. Computers do not directly calculate these partial differential derivatives as explained in Section 2.1.4. The partial derivative operators are replaced with basic operators, such as the four operations of arithmetic, which can be executed directly on a computer. The details are explained as follows.

2.2.4.1 Gradient

Nabla ∇ is expressed by components as follows:

$$\nabla = \begin{pmatrix} \dfrac{\partial}{\partial x} \\[2mm] \dfrac{\partial}{\partial y} \\[2mm] \dfrac{\partial}{\partial z} \end{pmatrix} \tag{2.22}$$

Thus, the nabla is expressed as a vector composed of the first partial derivatives of x, y, and z directions. We can calculate the gradient of an arbitrary function at a certain place by applying the nabla to the function at the position as follows:

$$\nabla \phi = \begin{pmatrix} \dfrac{\partial}{\partial x} \\[2mm] \dfrac{\partial}{\partial y} \\[2mm] \dfrac{\partial}{\partial z} \end{pmatrix} \phi = \begin{pmatrix} \dfrac{\partial \phi}{\partial x} \\[2mm] \dfrac{\partial \phi}{\partial y} \\[2mm] \dfrac{\partial \phi}{\partial z} \end{pmatrix} \tag{2.23}$$

Elements of the above vector, $\frac{\partial \phi}{\partial x}, \frac{\partial \phi}{\partial y}$, and $\frac{\partial \phi}{\partial z}$, express the gradients in x, y, and z directions, respectively. For example, if $\frac{\partial \phi}{\partial x}$ is 2 at a certain place, ϕ has a slope of 2 in the x-direction at the position. The slope of 2 means that if we increase x by infinitesimal length Δx, ϕ increases by $2\Delta x$. The function ϕ is an arbitrary quantity distributed spatially, such as pressure. The absolute value of the gradient vector expresses the steepness of the gradient at the position. The direction of the gradient vector expresses the direction in which the quantity increases most at the position.

2.2.4.2 The Gradient Model of the MPS Method (Nabla Model)

Computers do not calculate partial derivatives analytically. Therefore, we need to express the partial derivative operators using basic operators, such as the four basic operations of arithmetic (addition, subtraction, multiplication, and division). In the MPS method, we approximate $\nabla \phi$, which is the gradient of the arbitrary quantity, by the following model (Koshizuka and Oka, 1996):

$$\langle \nabla \phi \rangle_i = \frac{d}{n^0} \sum_{j \neq i} \left[\frac{\phi_j - \phi_i}{\left| \mathbf{r}_j - \mathbf{r}_i \right|^2} \left(\mathbf{r}_j - \mathbf{r}_i \right) w \left(\left| \mathbf{r}_j - \mathbf{r}_i \right|, r_e \right) \right] \qquad (2.24)$$

Eq. (2.24) is called the gradient model of the MPS method. The quantity ϕ is an arbitrary physical quantity, such as pressure. The parameter d is the number of spatial dimensions, and is 1, 2, or 3 for one-, two- or three-dimensional simulations, respectively. The angle brackets $\langle \rangle_i$ indicate that the operator in the brackets is approximated by a model of particles at the position of the ith particle. We find that the gradient of a quantity, which is expressed by the first partial derivatives in x, y, and z directions, shown in the left-hand side of Eq. (2.24) is approximated by the right-hand side, which is expressed by basic operators without partial differential operators. Computers can easily calculate the right-hand side of Eq. (2.24) and can obtain the approximate value of $\nabla \phi$ at the i th particle position.

2.2.4.3 The Meaning of Each Parts of the Gradient Model

Let us consider a specific case where a pressure P is applied as the arbitrary quantity ϕ below. Then, the pressure gradient is expressed as follows:

$$\langle \nabla P \rangle_i = \frac{d}{n^0} \sum_{j \neq i} \left[\frac{P_j - P_i}{\left| \mathbf{r}_j - \mathbf{r}_i \right|^2} \left(\mathbf{r}_j - \mathbf{r}_i \right) w \left(\left| \mathbf{r}_j - \mathbf{r}_i \right|, r_e \right) \right] \qquad (2.25)$$

The above model is easier to understand by expressing it as follows:

$$\langle \nabla P \rangle_i = \frac{d}{n^0} \sum_{j \neq i} \left[\frac{\left(P_j - P_i \right)}{\left| \mathbf{r}_j - \mathbf{r}_i \right|} \frac{\left(\mathbf{r}_j - \mathbf{r}_i \right)}{\left| \mathbf{r}_j - \mathbf{r}_i \right|} w \left(\left| \mathbf{r}_j - \mathbf{r}_i \right|, r_e \right) \right] \qquad (2.26)$$

The conceptual image of the gradient model is depicted in Fig. 2.5. In this figure, the bar on each particle expresses the height of the particle's pressure. The part $\frac{\left(P_j - P_i \right)}{\left| \mathbf{r}_j - \mathbf{r}_i \right|}$ in Eq. (2.26) expresses the steepness of the pressure gradient in the direction of $\mathbf{r}_j - \mathbf{r}_i$ at the ith particle position because the pressure difference between the ith and jth particles is divided by the distance between the two particles. The part $\frac{\left(\mathbf{r}_j - \mathbf{r}_i \right)}{\left| \mathbf{r}_j - \mathbf{r}_i \right|}$ in the right-hand side of Eq. (2.26) expresses the unit direction vector (the absolute value of the direction vector is 1) directed from the ith particle to the jth particle.

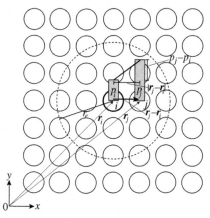

Figure 2.5 Schematic diagram of the gradient model of the MPS method in two dimensions.

Only $\left(\mathbf{r}_j - \mathbf{r}_i\right)$ is a vector, which has x-, y-, and z-directional components. The other parts in the gradient model are scalar values.

The function $w\left(\left|\mathbf{r}_j - \mathbf{r}_i\right|, r_e\right)$ on the right-hand side of Eq. (2.25) is a weight function, and we calculate it using Eq. (2.18). The brackets of $w\left(\left|\mathbf{r}_j - \mathbf{r}_i\right|, r_e\right)$ indicate that the arguments of the weight function $w(\ )$ are $\left|\mathbf{r}_j - \mathbf{r}_i\right|$ and r_e (they do not express the multiplication of w and $\left(\left|\mathbf{r}_j - \mathbf{r}_i\right|, r_e\right)$). That is, the weight function is calculated from the parameters $\left|\mathbf{r}_j - \mathbf{r}_i\right|$ and r_e. For details, see Section 2.2.3.

Let us review Eq. (2.25). We find that the gradient model of the MPS method calculates the slope of physical quantity between particle i and a neighboring particle j. As there are many neighboring particles, the gradient model calculates the weighted average of slopes between particles by using summation $\sum$ and $1/n^0$.

Note 2.13: *Reason why $1/n^0$ is necessary in the gradient model*: In Eq. (2.25), the result of the summation is divided by the standard particle number density n^0 to normalize the total weight to obtain the weighted average. This normalization is compared to the calculation of the average mark of an examination in a class. If there are 10 students in a class, you sum all the marks of students and divide the total by 10. The division by 10 corresponds to the division by n^0 in the gradient model (Eq. 2.25). In the calculation of the average mark, we apply a constant weight of 1. On the other hand, in the MPS calculation, we apply a weight function which depends on the distance between particles. Using the weight function, which depends on the distance, we can evaluate the gradient of a small region because we can assign a large influence (a weight) to a neighboring particle located near the point of interest.

Note 2.14: *Reason why d is necessary for the gradient model*: The reason why we need to multiply the number of spatial dimensions, d, in Eq. (2.25) is that the gradient will decrease to about $1/d$ times of the exact value if the above-mentioned weighted average is simply taken without d. The gradient model of the MPS method is expressed as a weighted average of slopes in the direction of each neighborhood particle, as expressed by Eq. (2.26). Because gradients are vectors, in addition to the information of slope in the direction from the ith particle to a neighboring particle j, we need information about the components of the slope in orthogonal directions. This is because a vector has three components in three-dimensional calculations, and it has two components in two-dimensional calculation. If the gradient model in a two-dimensional calculation is expressed without using the coefficient "d," Eq. (2.25) would be written as follows:

$$\langle \nabla \phi \rangle_i = \frac{1}{n^0} \sum_{j \neq i} \left[\left\{ \frac{\left(\phi_j - \phi_i \right) \left(\mathbf{r}_j - \mathbf{r}_i \right)}{\left| \mathbf{r}_j - \mathbf{r}_i \right| \left| \mathbf{r}_j - \mathbf{r}_i \right|} + \frac{\left(\phi_{j_\perp} - \phi_i \right) \left(\mathbf{r}_{j_\perp} - \mathbf{r}_i \right)}{\left| \mathbf{r}_{j_\perp} - \mathbf{r}_i \right| \left| \mathbf{r}_{j_\perp} - \mathbf{r}_i \right|} \right\} w \left(\left| \mathbf{r}_j - \mathbf{r}_i \right|, r_e \right) \right]$$

(2.27)

where $\phi_{j_\perp}$ is the physical quantity of $j_\perp$th particle located at $\mathbf{r}_{j_\perp}$. Fig. 2.6 shows the schematic diagram of Eq. (2.27). The definition of $\mathbf{r}_{j_\perp}$ is as follows:

$$\mathbf{r}_{j_\perp} \equiv \mathbf{r}_i + \left| \mathbf{r}_j - \mathbf{r}_i \right| \mathbf{n}_{ij_\perp}$$

(2.28)

$$\mathbf{n}_{ij_\perp} \cdot \left(\mathbf{r}_j - \mathbf{r}_i \right) = 0$$

(2.29)

$$\left| \mathbf{n}_{ij_\perp} \right| = 1$$

(2.30)

where $\mathbf{n}_{ij_\perp}$ is a unit vector which is at a right angle to $\left(\mathbf{r}_j - \mathbf{r}_i \right)$ as expressed in Eqs. (2.29) and (2.30). The vectors $\mathbf{r}_{j_\perp}$ and $\mathbf{n}_{ij_\perp}$ are used only for the explanation of parameter "d" in Eq. (2.25), actually we do not need to calculate nor use them. By Eq. (2.27), we can consider the

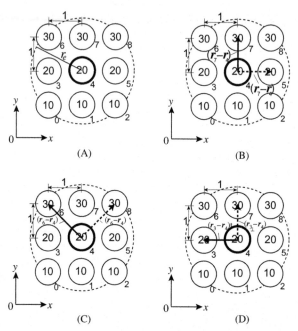

Figure 2.6 Schematic diagram indicating the necessity of d in the gradient model of the MPS method: (A) Neighboring particles around the 4th particle, (B) the case where the 7th particle is the j-th particle, (C) the case where the 6th particle is the j-th particle, and (D) the case where the 3rd particle is the j-th particle. *The value in each particle expresses the pressure of the particle. The numbers in inferior letters are IDs of particles.*

components of slope in two orthogonal directions simultaneously. Therefore, we can calculate the gradient in two dimensions. In a three-dimensional calculation, by adding another orthogonal component, we can calculate the gradient in the same manner. However, it is difficult to find particle $j_\perp$ among neighboring particles around particle i. Every neighboring particle j can become $j_\perp$ two or three times in two or three dimensions, respectively, because we assume that particle distribution is uniform and symmetric around the ith particle. Therefore, we multiply by "d" in Eq. (2.25) for calculating gradients instead of using Eq. (2.27).

Although we assume that neighboring particles are distributed symmetrically around particle i, and there is a particle $j_\perp$ around particle i, actually the particle distribution is not always symmetric, and particle $j_\perp$ does not always exist. In an asymmetric particle distribution, a numerical error due to the assumption of the particle distribution occurs. The MPS method introduces the assumption of uniform and symmetrical particle distribution by using an adequate effective radius of interaction zone. To reduce the error due to particle distribution, new models have been developed. For details, see Tamai and Koshizuka (2014).

2.2.4.4 Example of Gradient Calculation

Let us calculate a gradient by the gradient model of the MPS method to confirm our understanding of the gradient model. Although we calculate gradients with a computer usually, we will calculate with a paper, pencil, and scientific electronic calculator to deepen our understanding. We assume a two-dimensional fluid which has a pressure distribution. The fluid is expressed by fluid particles as shown in Fig. 2.7. The value written in each particle is the pressure of the particle. The unit of pressure is Pascals. The number written at the lower right of each particle is the particle ID. We calculate the pressure gradient at the position of the particle whose ID is 12 in two dimensions. The initial and current distances between particles are set to 1 m. The effective radius of the interaction zone is set to $1.5l_0$. Because the pressure distribution and the particle distribution are very simple, we find that the exact solution of the pressure gradient is $(0, \ 10)^T$, which means that the slope of the pressure is 0 and 10 Pa/m in x- and y-direction, respectively. Let us verify whether the gradient model of Eq. (2.25) can obtain $(0, \ 10)^T$ below. (The weights of the particles whose IDs are 0−5, 9, 10, 14, 15, and 19−24 become zero because the distance from the 12th particle is longer than r_e. Therefore, the terms for these particles were disregarded in Eq. 2.31).

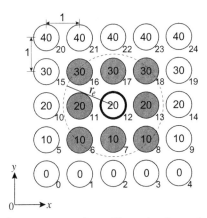

Figure 2.7 Example of pressure gradient. *The value in each particle expresses the pressure of the particle. The numbers in inferior letters are IDs of particles.*

$$\langle \nabla P \rangle_{12} = \frac{d}{n^0} \sum_{j \neq 12} \left[\frac{P_j - P_{12}}{|\mathbf{r}_j - \mathbf{r}_{12}|^2} (\mathbf{r}_j - \mathbf{r}_{12}) w(|\mathbf{r}_j - \mathbf{r}_{12}|, r_e) \right]$$

$$= \frac{2}{n^0} \times \sum_{j \neq 12} \left[\frac{P_j - P_{12}}{|\mathbf{r}_j - \mathbf{r}_{12}|^2} (\mathbf{r}_j - \mathbf{r}_{12}) w(|\mathbf{r}_j - \mathbf{r}_{12}|, r_e) \right]$$

$$= \frac{2}{n^0} \times \left\{ \frac{P_6 - P_{12}}{|\mathbf{r}_6 - \mathbf{r}_{12}|^2} (\mathbf{r}_6 - \mathbf{r}_{12}) w(|\mathbf{r}_6 - \mathbf{r}_{12}|, r_e) + \frac{P_7 - P_{12}}{|\mathbf{r}_7 - \mathbf{r}_{12}|^2} (\mathbf{r}_7 - \mathbf{r}_{12}) w(|\mathbf{r}_7 - \mathbf{r}_{12}|, r_e) \right.$$

$$+ \frac{P_8 - P_{12}}{|\mathbf{r}_8 - \mathbf{r}_i|^2} (\mathbf{r}_8 - \mathbf{r}_{12}) w(|\mathbf{r}_8 - \mathbf{r}_{12}|, r_e) + \frac{P_{11} - P_{12}}{|\mathbf{r}_{11} - \mathbf{r}_{12}|^2} (\mathbf{r}_{11} - \mathbf{r}_{12}) w(|\mathbf{r}_{11} - \mathbf{r}_{12}|, r_e)$$

$$+ \frac{P_{13} - P_{12}}{|\mathbf{r}_{13} - \mathbf{r}_{12}|^2} (\mathbf{r}_{13} - \mathbf{r}_{12}) w(|\mathbf{r}_{13} - \mathbf{r}_{12}|, r_e) + \frac{P_{16} - P_{12}}{|\mathbf{r}_{16} - \mathbf{r}_{12}|^2} (\mathbf{r}_{16} - \mathbf{r}_{12}) w(|\mathbf{r}_{16} - \mathbf{r}_{12}|, r_e)$$

$$+ \frac{P_{17} - P_{12}}{|\mathbf{r}_{17} - \mathbf{r}_{12}|^2} (\mathbf{r}_{17} - \mathbf{r}_{12}) w(|\mathbf{r}_{17} - \mathbf{r}_{12}|, r_e) + \left. \frac{P_{18} - P_{12}}{|\mathbf{r}_{18} - \mathbf{r}_{12}|^2} (\mathbf{r}_{18} - \mathbf{r}_{12}) w(|\mathbf{r}_{18} - \mathbf{r}_{12}|, r_e) \right\}$$

$$= \frac{2}{n^0} \times \left\{ \frac{10 - 20}{\sqrt{2}^2} \times ((2,2)^T - (3,3)^T) \times w(\sqrt{2}, 1.5) + \frac{10 - 20}{1^2} \right.$$

$$\times ((3,2)^T - (3,3)^T) \times w(1, 1.5) + \frac{10 - 20}{\sqrt{2}^2} \times ((4,2)^T - (3,3)^T) \times w(\sqrt{2}, 1.5)$$

$$+ \frac{20 - 20}{1^2} \times ((2,3)^T - (3,3)^T) \times w(1, 1.5) + \frac{20 - 20}{1^2} \times ((4,3)^T - (3,3)^T)$$

$$\times w(1, 1.5) + \frac{30 - 20}{\sqrt{2}^2} \times ((2,4)^T - (3,3)^T) \times w(\sqrt{2}, 1.5) + \frac{30 - 20}{1^2}$$

$$\times ((3,4)^T - (3,3)^T) \times w(1, 1.5) + \left. \frac{30 - 20}{\sqrt{2}^2} \times ((4,4)^T - (3,3)^T) \times w(\sqrt{2}, 1.5) \right\}$$

$$= \frac{2}{n^0} \times \left\{ -5 \times (-1,-1)^T \times w(\sqrt{2}, 1.5) - 10 \times (0,-1)^T \times w(1, 1.5) - 5 \times (1,-1)^T \right.$$

$$\times w(\sqrt{2}, 1.5) 5 \times (-1,1)^T \times w(\sqrt{2}, 1.5) + 10 \times (0,1)^T \times w(1, 1.5) + 5 \times (1,1)^T \times w(\sqrt{2}, 1.5) \left. \right\}$$

$$= \frac{2}{4 \times w(\sqrt{2}, 1.5) + 4 \times w(1, 1.5)} \times \left\{ (0,20)^T \times w(\sqrt{2}, 1.5) + (0,20)^T \times w(1, 1.5) \right\}$$

$$= \frac{2}{4 \times \left\{ w(\sqrt{2}, 1.5) + w(1, 1.5) \right\}} \times (0,5)^T \times 4 \times \left\{ w(\sqrt{2}, 1.5) + w(1, 1.5) \right\} = 2 \times (0,5)^T = (0,10)^T$$

$$(2.31)$$

By the above calculation, it was verified that the gradient model of the MPS method expressed in Eq. (2.25) can accurately calculate the pressure gradient.

2.2.4.5 Laplacian Operator and Its Uses

Laplacian operator ∇^2 is the differential operator expressed as follows:

$$\nabla^2\phi = \nabla\cdot\nabla\phi = \begin{pmatrix} \dfrac{\partial}{\partial x} \\[2mm] \dfrac{\partial}{\partial y} \\[2mm] \dfrac{\partial}{\partial z} \end{pmatrix} \cdot \begin{pmatrix} \dfrac{\partial}{\partial x} \\[2mm] \dfrac{\partial}{\partial y} \\[2mm] \dfrac{\partial}{\partial z} \end{pmatrix}\phi = \left(\dfrac{\partial^2}{\partial x^2} + \dfrac{\partial^2}{\partial y^2} + \dfrac{\partial^2}{\partial z^2}\right)\phi$$

$$= \dfrac{\partial^2\phi}{\partial x^2} + \dfrac{\partial^2\phi}{\partial y^2} + \dfrac{\partial^2\phi}{\partial z^2}$$

$$(2.32)$$

From the right-hand side of Eq. (2.32), we find that the Laplacian of a certain quantity ϕ is expressed as the sum of the second derivatives in x, y, and z directions. Laplacian operators appear in the viscosity term of the Navier−Stokes equations and the pressure Poisson equation.

2.2.4.6 The Laplacian Model of the MPS Method

We discretize the Laplacian operator by the following Laplacian model (Koshizuka and Oka, 1996):

$$\langle\nabla^2\phi\rangle_i = \frac{2d}{\lambda^0 n^0}\sum_{j\neq i}\left(\phi_j - \phi_i\right)w\left(\left|\mathbf{r}_j - \mathbf{r}_i\right|, r_e\right) \qquad (2.33)$$

The parameter d in Eq. (2.3) is the number of spatial dimensions. That is, d is 2 in two-dimensional simulations, and is 3 in three-dimensional simulations. The division by n^0 in Eq. (2.33) is normalizes the total weight in the same manner as the gradient model. The parameter λ^0 expresses the average of the squared distance among particles and is calculated as follows:

$$\lambda^0 = \frac{\displaystyle\sum_{j\neq i'}\left|\mathbf{r}_j^0 - \mathbf{r}_{i'}^0\right|^2 w\left(\left|\mathbf{r}_j^0 - \mathbf{r}_{i'}^0\right|, r_e\right)}{\displaystyle\sum_{j\neq i'}w\left(\left|\mathbf{r}_j^0 - \mathbf{r}_{i'}^0\right|, r_e\right)} \qquad (2.34)$$

where i' is a particle located in the inside of the fluid at the initial state of a simulation, where particles are arranged at equal intervals. We use the calculated λ^0 as a common constant value among all the particles over a simulation.

The coefficients $\frac{2d}{\lambda^0 n^0}$ are necessary for calibrating the absolute value of the Laplacian. For details about the derivation of these coefficients, see Section 1.2.2.

As shown in Eq. (2.33), the Laplacian operator on the left-hand side is expressed without partial derivatives on the right-hand side. Therefore, computers can easily calculate the Laplacian by the calculation of the right-hand side. The MPS method expresses Laplacian operators without considering the second derivatives in x, y, and z directions. We explain the verification of the Laplacian in Section 2.2.4.6.2. Moreover, the model of Eq. (2.33) is symmetrical between particles i and j. Therefore, if we discretize the Laplacian operator of the pressure Poisson equation by the Laplacian model, the coefficient matrix becomes symmetric, and a fast solver can be applied to solve the simultaneous equations of the discretized pressure Poisson equation. Moreover, if we apply the Laplacian model of the MPS method to the viscosity term of the Navier—Stokes equation, the momentum is conserved between particles in the viscosity calculation. These are advantages inherent in the Laplacian model of the MPS method.

2.2.4.6.1 The Meaning of the Laplacian Model

The Laplacian operator expresses diffusion of a certain quantity ϕ because it appears in diffusion equations. For details, see Note 2.10 in Section 2.2.2.1. A schematic diagram of the Laplacian model in a

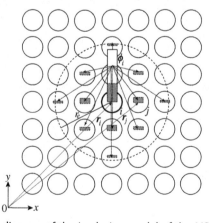

Figure 2.8 Schematic diagram of the Laplacian model of the MPS method.

two-dimensional simulation is shown in Fig. 2.8. This figure expresses the case where the Laplacian operator is applied to a diffusion equation, and a physical quantity of a certain particle i is being distributed to the neighboring particles j by the Laplacian model of the MPS method. The bar on particle i is the physical quantity which the particle originally has. The term $-\phi_i w(|\mathbf{r}_j - \mathbf{r}_i|, r_e)$ in Eq. (2.33) expresses that the physical quantity of particle i is distributed to a neighboring particle j in proportion to the physical quantity of particle i and the weight function. The weight function is a function expressed by the distance between particles as explained in Section 2.2.3, and becomes large if the distance between particles is short. As a result, the physical quantity to be distributed from particle i to particle j becomes larger as the distance between particles becomes shorter. The bar on a neighboring particle j expresses the physical quantity that particle j received from particle i. We can explain the reason why $-\phi_i w(|\mathbf{r}_j - \mathbf{r}_i|, r_e)$ has a negative sign. The reason is that the physical quantity of particle i is reduced because of the diffusion. The white region in the bar on particle i in Fig. 2.8 expresses the quantity lost by the distribution from particle i to its neighboring particles. The distribution also occurs from particle j to particle i, and is expressed by the term $\phi_j w(|\mathbf{r}_j - \mathbf{r}_i|, r_e)$ in Eq. (2.33). We can calculate the Laplacian in three dimensions in the same manner as in two-dimensional simulations.

2.2.4.6.2 Example of Laplacian Calculation

Consider the one-dimensional case shown in Fig. 2.9 to confirm our understanding of the Laplacian model. We calculate the Laplacian of the physical quantity ϕ at the position of the particle i. The particle i has two neighboring particles whose IDs are $i+1$ and $i-1$. The distance l_0, which expresses the distance between particles at the initial state, is set to Δx. The effective radius of the interaction zone, r_e, is set to $1.5l_0$. To simplify the problem, we assume all particles are stationary. First, we calculate λ^0 from Eq. (2.34) to calculate the Laplacian as follows:

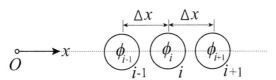

Figure 2.9 Particle arrangement for understanding the Laplacian model of the MPS method in one dimension.

$$\lambda^0 = \frac{\left|\mathbf{r}_{i+1}^0 - \mathbf{r}_i^0\right|^2 w\left(\left|\mathbf{r}_{i+1}^0 - \mathbf{r}_i^0\right|, r_e\right) + \left|\mathbf{r}_{i-1}^0 - \mathbf{r}_i^0\right|^2 w\left(\left|\mathbf{r}_{i-1}^0 - \mathbf{r}_i^0\right|, r_e\right)}{w\left(\left|\mathbf{r}_{i+1}^0 - \mathbf{r}_i^0\right|, r_e\right) + w\left(\left|\mathbf{r}_{i-1}^0 - \mathbf{r}_i^0\right|, r_e\right)}$$

$$= \frac{(\Delta x)^2 w(\Delta x, r_e) + (\Delta x)^2 w(\Delta x, r_e)}{w(\Delta x, r_e) + w(\Delta x, r_e)} = \frac{2(\Delta x)^2 w(\Delta x, r_e)}{2w(\Delta x, r_e)} = (\Delta x)^2$$

$$(2.35)$$

Then, we use the calculated λ^0 for the Laplacian model of Eq. (2.33). As the simulation is one-dimensional, d is 1. As a result, the Laplacian of an arbitrary quantity of particle i is calculated as follows:

$$\left\langle \nabla^2 \phi \right\rangle_i = \frac{2d}{\lambda^0 n^0} \sum_{j \neq i} \left(\phi_j - \phi_i\right) w\left(\left|\mathbf{r}_j - \mathbf{r}_i\right|, r_e\right)$$

$$= \frac{2 \cdot 1}{(\Delta x)^2 \cdot \left\{ w\left(\left|\mathbf{r}_{i+1}^0 - \mathbf{r}_i^0\right|, r_e\right) + w\left(\left|\mathbf{r}_{i-1}^0 - \mathbf{r}_i^0\right|, r_e\right) \right\}}$$
$$\left\{ \left(\phi_{i+1} - \phi_i\right) w(|\mathbf{r}_{i+1} - \mathbf{r}_i|, r_e) + \left(\phi_{i-1} - \phi_i\right) w(|\mathbf{r}_{i-1} - \mathbf{r}_i|, r_e) \right\}$$

$$= \frac{2}{(\Delta x)^2 \cdot \left\{ w(\Delta x, r_e) + w(\Delta x, r_e) \right\}}$$
$$\left\{ \left(\phi_{i+1} - \phi_i\right) w(\Delta x, r_e) + \left(\phi_{i-1} - \phi_i\right) w(\Delta x, r_e) \right\}$$

$$= \frac{2}{(\Delta x)^2 \cdot 2w(\Delta x, r_e)} \left[\left(\phi_{i+1} - \phi_i\right) + \left(\phi_{i-1} - \phi_i\right)\right] w(\Delta x, r_e)$$

$$= \frac{\phi_{i+1} - 2\phi_i + \phi_{i-1}}{(\Delta x)^2}$$

$$(2.36)$$

Because the above example is one-dimensional, $\left\langle \nabla^2 \phi \right\rangle_i$ expresses $\left\langle \frac{\partial^2}{\partial x^2} \phi \right\rangle_i$, which is the discretized $\frac{\partial^2 \phi}{\partial x^2}$ at the position of the ith particle. We thus find that the right-hand side of Eq. (2.36) is in agreement with that of the finite difference method expressed as follows:

$$\frac{\partial^2 \phi}{\partial x^2} \cong \frac{\frac{(\phi_{i+1} - \phi_i)}{\Delta x} - \frac{(\phi_i - \phi_{i-1})}{\Delta x}}{(\Delta x)} = \frac{\phi_{i+1} - 2\phi_i + \phi_{i-1}}{(\Delta x)^2} \qquad (2.37)$$

Moreover, we find that λ^0 of the MPS method corresponds to $(\Delta x)^2$ of the finite difference method. The denominator $(\Delta x)^2$ is necessary for the Laplacian model because the Laplacian operator consists of second derivatives with respect to space. It is necessary to divide a quantity by a square of minute length.

Let us calculate a second derivative by the Laplacian model, as expressed in Eq. (2.36), to confirm our understanding. We consider a one-dimensional calculation where there are three particles whose IDs are $i + 1$, i, and $i - 1$, as shown in Fig. 2.9. We calculate the second derivative of ϕ at the position of particle i. We assume each particle has a physical quantity expressed by a parabola as follows:

$$\phi = a(x - x_i)^2 + b \tag{2.38}$$

where x is the x-coordinate of an arbitrary particle, and x_i is the x-coordinate of particle i, which is located at the center of the particles. The parameters a and b are arbitrary constants. By Eq. (2.38), the physical quantities of the particles $i + 1$, i, and $i - 1$ are set to $a \times (\Delta x)^2 + b$, $a \times 0^2 + b$, and $a \times (-\Delta x)^2 + b$, respectively. Note that the setting of the physical quantity based on Eq. (2.38) is only for explanation of Laplacian and is not necessary for an actual MPS simulation. The Laplacian of ϕ at the position of particle i is calculated by Eq. (2.36) as follows:

$$
\begin{aligned}
\langle \nabla^2 \phi \rangle_i &= \frac{\phi_{i+1} - 2\phi_i + \phi_{i-1}}{\Delta x^2} \\
&= \frac{\left\{ a \times (\Delta x)^2 + b \right\} - 2 \times \left(a \times 0^2 + b \right) + \left\{ a \times (-\Delta x)^2 + b \right\}}{\Delta x^2} \\
&= \frac{2 \left\{ a \times (\Delta x)^2 \right\}}{\Delta x^2} = 2a
\end{aligned}
\tag{2.39}
$$

We find that the second derivative of $\phi = a(x - x_i)^2 + b$ can be calculated correctly in regular intervals. Note that, if the particle distribution is irregular, an error due to the irregular particle distribution occurs. To reduce this error, new models have been developed. For details, see Tamai and Koshizuka (2014).

2.2.5 Semi-implicit Method

2.2.5.1 How to Calculate Pressure, and the Necessity of the Semi-implicit Method

When calculating the acceleration of each particle by the Navier–Stokes equations at every time step, a problem arises. The problem is that we cannot obtain pressure from the Navier–Stokes equations. The Navier–Stokes equations express the time change of fluid velocity and do not express the time change of pressure. By integrating the Navier–Stokes equations by a program, such as Program 2.3 in Section 2.1.3, we can update the fluid velocity $\mathbf{u}^k$ to $\mathbf{u}^{k+1}$, where $\mathbf{u}^k$ and $\mathbf{u}^{k+1}$ are the fluid velocities of a particle at time step k and $k+1$, respectively. The velocity $\mathbf{u}^k$ corresponds to "velocity" on the right-hand side of Program 2.3. The velocity $\mathbf{u}^{k+1}$ corresponds to "velocity" on the left-hand side of Program 2.3. If we can obtain the physical quantity of the next time step by integrating an equation with respect to time, the equation is in the form of time development. The Navier–Stokes equation is in a form of time development with respect to velocity. However, it is not in a form of time development with respect to pressure. Therefore, we cannot update the pressure distribution using only the Navier–Stokes equations. Another equation is necessary to obtain pressure distribution for the next time step. The required equation is the equation of state or the equation of continuity.

Gas, which is a type of fluid, is easy to compress. This is because the distance between molecules is relatively long, and there is much room for compressing. If fluid is assumed as an ideal gas under an adiabatic condition, where there is no exchange of heat with the exterior, we can easily calculate pressure on the basis of the equation of state.

On the other hand, liquid is difficult to compress because the distance between molecules is very short. Molecules exist densely in a liquid. There is little room to be compressed. We can illustrate the above issue through an example of fluid in a cylinder with a piston. If we fill a cylinder with liquid, it is difficult to compress the liquid, while it is very easy to compress gas in a cylinder with a piston. The characteristic of the compressibility of liquid, which is very hard against compression, is difficult to express by the above-mentioned equation of state because we need to use a very small time increment Δt to express the hardness about the compression. (However, there are some methods which calculate the pressure of liquid on the basis of the equation of state by allowing a small amount of compression. For details, refer to Section 3.2.) If Δt is very small, a large number of time steps are required for simulating a phenomenon requiring long computation time.

Therefore, in most fluid simulations, liquid flows are treated as incompressible flows, which do not compress at all. This technique does not use the equation of state but uses the equation of continuity as a condition equation for calculating pressure. Although actual fluid can expand and contract very slightly (e.g. sound waves in water), the influences of the volume change are small enough and can be disregarded in many cases of liquid simulation. Moreover, in some cases of gas flows, if the flow velocities are very slow compared to the speed of sound, the flows can be approximated as incompressible flows. By applying the approximation of incompressibility to the equation of continuity (Eq. 2.7), and applying the approximated equation of continuity to the Navier−Stokes equations, we can delete some terms involving velocity and can calculate the pressure. This method calculates pressure by solving simultaneous linear equations which are expressed by unknown fluid pressures of the next time step and is called an "implicit method." On the other hand, the methods that calculate unknown values without solving simultaneous equations on the basis of the values at the current time step, are called "explicit methods." Implicit methods are able to use a relatively large time increment Δt. That is an advantage of implicit methods. However, implicit methods will take a relatively long time to solve the simultaneous linear equation. Therefore, computational time per time step of the implicit method is longer than that of the explicit method, although we can reduce the number of time steps by the large time increment of the implicit method. Therefore, we need to choose an implicit or explicit method considering these characteristics.

2.2.5.2 Outline of the Semi-implicit Method in the MPS Method

Let us study how to move fluid particles on the basis of a semi-implicit method. Fig. 2.10 shows a schematic diagram of particle motion for a time step in a dam breaking simulation, where a water column collapses because of gravity. Fig. 2.10A expresses the particle distribution at the initial state. Particles are arranged at regular intervals. As explained in Section 2.2.2, particles are moved on the basis of the acceleration terms on the right-hand side of the Navier−Stokes equations (Eq. 2.6). As the pressure of particles is an unknown, we first calculate the terms on the right-hand side of Eq. (2.6) other than the pressure gradient term. If the viscous force is disregarded to simplify the problem, then except for the pressure gradient term, only gravity influences fluid particles. As a result, the fluid particles have downward acceleration due to gravity. We accelerate the fluid-particle velocities downward at the acceleration of gravity and move particles at their accelerated velocities. The procedure for the particle displacement is the same as that in Programs 2.3 and 2.4 in Section 2.1.3. The phase just after the displacement is shown in Fig. 2.10B. Because the pressure gradient term has not been included yet

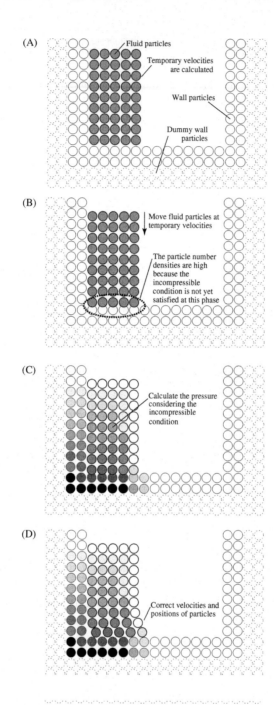

Figure 2.10 Schematic diagram of particle motion during a time step using the semi-implicit algorithm of the MPS method. (A) Initial phase of a time step, (B) the phase just after the particles' motion based on the temporary velocities, (C) Pressure distribution of particles, and (D) the phase after correcting particle velocities and positions by pressure gradient.

in this phase, there is an area where the densities of particles are not constant. That is, the condition of incompressibility is not satisfied yet. Then, we calculate pressure considering incompressibility. Fig. 2.10C shows a conceptual image of the pressure distribution. Then, we calculate $-\frac{1}{\rho}\nabla P$, which is the pressure gradient term using the calculated pressure distribution, and correct particle velocities and positions with the acceleration of the pressure gradient. The conceptual image of the particle distribution just after the correction is shown in Fig. 2.10D. Thus, (A) first we accelerate particle velocities by the acceleration of the Navier–Stokes equation other than the pressure gradient term, (B) second move the particles at the updated velocity, (C) then calculate pressure considering the condition of incompressibility, and (D) last we correct the velocities and positions of particles using the acceleration of the pressure gradient. The points (A)–(D) are necessary procedures for a time step in the MPS method. We repeat the series of procedures until the simulation time reaches a designated completion time.

2.2.5.3 Details of the Semi-implicit Method of the MPS Method

The semi-implicit method of the MPS method solves the Navier–Stokes equations on the basis of the fractional step method, which is called the SMAC method and is often used in grid-based methods. Let us review the Navier–Stokes equations, which are the equations of motion for fluid. The incompressible Navier–Stokes equations are expressed as follows:

$$\frac{D\mathbf{u}}{Dt} = -\frac{\nabla P}{\rho^0} + \nu\nabla^2\mathbf{u} + \mathbf{g} \qquad (2.40)$$

where the fluid density ρ is expressed as ρ^0. The superior letter 0 expresses the value at the initial time step. In the semi-implicit algorithm of the MPS method, fluid flows are treated as incompressible flows. Therefore, ρ is assumed to be a constant ρ^0. The superior letter 0 is written to distinguish the fluid density from ρ^*, which is the fluid density at a phase before the incompressible condition is imposed. The details of ρ^* are explained later.

In general, computers do not analytically calculate $\frac{D}{Dt}$, ∇, and ∇^2, which are operators involving differentiation. In the MPS method, ∇ and ∇^2 are replaced by $\langle\nabla\rangle$ and $\langle\nabla^2\rangle$, which are the particle approximation models explained in Section 2.2.4. The term $\frac{D\mathbf{u}_i}{Dt}$ on the left-hand side of Eq. (2.40) expresses the time change of velocity vector $\mathbf{u}_i$ of a fluid particle i. The operator $\frac{D}{Dt}$ is called the "material derivative," "Lagrangian derivative," or "substantial derivative," and expresses the time change of a quantity of a material point. In order to integrate $\frac{D\mathbf{u}_i}{Dt}$ and obtain the velocity at the next time step using computers, $\frac{D\mathbf{u}_i}{Dt}$ is discretized as follows:

$$\frac{D\mathbf{u}_i}{Dt} \cong \frac{\mathbf{u}_i^{k+1} - \mathbf{u}_i^k}{\Delta t} \tag{2.41}$$

The vectors $\mathbf{u}_i^k$ and $\mathbf{u}_i^{k+1}$ are the velocities of the ith particle at the k- and $(k+1)$th time steps, respectively. The scalar parameter Δt is the time increment of a simulation and is defined as follows:

$$\Delta t \equiv t^{k+1} - t^k \tag{2.42}$$

where t^k and t^{k+1} are times at kth and $(k+1)$th time steps, respectively. Parameters with superior letter k are known values because these parameters express values at the present time. On the other hand, parameters with the superior letter $k+1$ are unknown values because these parameters express future values. By replacing $\frac{D\mathbf{u}_i}{Dt}$ of Eq. (2.40) with $\frac{\mathbf{u}_i^{k+1} - \mathbf{u}_i^k}{\Delta t}$, we can express the material derivative using the four basic operations of arithmetic, which are easily calculated by computers. As a result, Eq. (2.40) is expressed as follows:

$$\frac{\mathbf{u}_i^{k+1} - \mathbf{u}_i^k}{\Delta t} = -\frac{1}{\rho^0}\langle\nabla P\rangle_i^{k+1} + \nu\langle\nabla^2\mathbf{u}\rangle_i^k + \mathbf{g} \tag{2.43}$$

The pressure P^{k+1} is the pressure at the next time and is an unknown value. Therefore, we cannot calculate all the terms on the right-hand side of Eq. (2.44) at once. Therefore, we first calculate Eq. (2.43) by neglecting the pressure gradient term and calculate the temporary velocity by the following equation:

$$\frac{\mathbf{u}_i^* - \mathbf{u}_i^k}{\Delta t} = \nu\langle\nabla^2\mathbf{u}\rangle_i^k + \mathbf{g} \tag{2.44}$$

where $\mathbf{u}_i^*$ is the temporal velocity, which is obtained by the Navier–stokes equation without the pressure gradient term. Specifically, we can calculate $\mathbf{u}_i^*$ by the following equation, which is derived by multiplying Δt on both sides of Eq. (2.44) and transposing $\mathbf{u}_i^k$ to the right-hand side as follows:

$$\mathbf{u}_i^* = \mathbf{u}_i^k + \left\{\nu\langle\nabla^2\mathbf{u}\rangle_i^k + \mathbf{g}\right\}\Delta t \tag{2.45}$$

All the terms on the right-hand side are expressed by known parameters. By calculating the right-hand side of Eq. (2.45), we can obtain the temporary velocity $\mathbf{u}_i^*$, which is the left-hand side of Eq. (2.45).

A calculation such as Eq. (2.45) is called "time integration by the forward Euler method." We calculate $\langle \nabla^2 \mathbf{u} \rangle_i^k$ by expressing it with the Laplacian model (Eq. 2.33) of the MPS method. As a result, Eq. (2.45) is expressed without symbol of the partial differential operator as follows:

$$\mathbf{u}_i^* = \mathbf{u}_i^k + \left\{ \nu \frac{2d}{\lambda^0 n^0} \sum_{j \neq i} \left(\mathbf{u}_j^k - \mathbf{u}_i^k \right) w \left(\left| \mathbf{r}_j^k - \mathbf{r}_i^k \right|, r_e \right) + \mathbf{g} \right\} \Delta t \qquad (2.46)$$

The parts shown in {} on the right-hand side of Eq. (2.46) express an acceleration vector. Therefore, we can calculate Eq. (2.46) by a similar program as Program 2.3 in Section 2.1.3. We calculate the right-hand side for all fluid particles and obtain the temporal velocities of all fluid particles. The state, where the calculation of temporary velocities is completed, corresponds to Fig. 2.10A.

Next, we move particles to their temporary positions $\mathbf{r}_i^*$ with the calculated temporary velocities $\mathbf{u}_i^*$ by the following equation:

$$\mathbf{r}_i^* = \mathbf{r}_i^k + \mathbf{u}_i^* \Delta t \qquad (2.47)$$

The state just after the above displacement corresponds to Fig. 2.10B. We can calculate Eq. (2.47) by the same program as Program 2.4 in Section 2.1.3. Eq. (2.47) is derived by an approximation of velocity on the basis of the forward Euler method as follows:

$$\mathbf{u}_i^* = \frac{D\mathbf{r}_i^*}{Dt} \cong \frac{\mathbf{r}_i^* - \mathbf{r}_i^k}{\Delta t} \qquad (2.48)$$

Specifically, Eq. (2.47) is derived by multiplying both sides of Eq. (2.48) by Δt, transposing the unknown value $\mathbf{r}_i^*$ to the left-hand side, transposing a known value $\mathbf{u}_i^*$ to the right-hand side, and finally multiplying (-1) to both sides.

Then, at the particle distribution of the temporal particle position $\mathbf{r}*$, we calculate the pressure distribution of particles. This corresponds to the state of Fig. 2.10C. Specifically, we calculate the pressure distribution by the pressure Poisson equation expressed as follows (Note 2.15):

$$\langle \nabla^2 P \rangle_i^{k+1} = -\rho^0 \frac{1}{(\Delta t)^2} \frac{n_i^* - n^0}{n^0} \qquad (2.49)$$

where n_i^* is the particle number density of the ith particle at the temporary particle position $\mathbf{r}_i^*$. Eq. (2.49) does not express the substitution of the right-hand side for the left-hand side, but expresses the relationship between fluid pressure on the left-hand side and the particle number density on the right-hand side. We then calculate the pressure distribution that satisfies Eq. (2.49). To calculate pressure distribution, we discretize the Laplacian of the left-hand side of Eq. (2.49) by the Laplacian model of the MPS method (Eq. 2.33), and express the Laplacian by the four operations of arithmetic without partial differential operators so that computers can calculate it. As a result, Eq. (2.49) is expressed by simultaneous linear equations whose unknown values are the pressure P^{k+1} of particles. The details of the derivation of Eq. (2.49) and how to solve the simultaneous equations are explained in Sections 2.2.5.4 and 2.2.5.5.

Note 2.15: *Poisson equation*: The Poisson equation is an elliptic differential equation expressed as $\nabla^2 \phi = b$, where ϕ is an arbitrary quantity. The parameter b is often called the "source term," and is a function or constant which is not expressed by ϕ.

Then, we calculate the true velocity vector $\mathbf{u}_i^{k+1}$ of the next time step using the calculated pressure P^{k+1} and the temporary velocity $\mathbf{u}_i^*$ by the following equation:

$$\mathbf{u}_i^{k+1} = \mathbf{u}_i^* - \frac{1}{\rho^0} \langle \nabla P \rangle_i^{k+1} \Delta t \qquad (2.50)$$

The above equation indicates that the true velocity of the next time step is calculated by adding the temporal velocity $\mathbf{u}_i^*$ to the velocity change due to the acceleration by the pressure gradient term. Eq. (2.50) is derived from the following equation:

$$\frac{\mathbf{u}_i^{k+1} - \mathbf{u}_i^*}{\Delta t} = -\frac{1}{\rho^0} \langle \nabla P \rangle_i^{k+1} \qquad (2.51)$$

For the details of Eq. (2.51), see Section 2.3.4.16, where the discretized version of Eq. (2.51) and the program to solve it are explained.

We can find that the terms of the temporary velocity $\mathbf{u}_i^*$ are canceled if we add Eqs. (2.44) and (2.51). As a result, the obtained equation is equal to the discretized Navier–Stokes equations given by Eq. (2.43). Therefore, we can find that the particle velocity updated by Eqs. (2.45) and (2.50) is the approximated solution of the Navier–Stokes equations.

Finally, we correct the position of particles from $\mathbf{r}_i^*$ to $\mathbf{r}_i^{k+1}$ by using the calculated $\mathbf{u}_i^{k+1}$ as follows:

$$\mathbf{r}_i^{k+1} = \mathbf{r}_i^* + \left(\mathbf{u}_i^{k+1} - \mathbf{u}_i^* \right) \Delta t \qquad (2.52)$$

where $\mathbf{r}_i^{k+1}$ is the true position of the ith particle of the next time step $k + 1$. We carry out these procedures for all particles. The phase where this processing has just finished, corresponds to Fig. 2.10D. We can calculate Eq. (2.52) in the same manner as Program 2.4 in Section 2.1.3. Eq. (2.52) is derived as follows:

$$
\begin{aligned}
\mathbf{r}_i^{k+1} &= \mathbf{r}_i^k + \mathbf{u}_i^{k+1}\Delta t = \mathbf{r}_i^k + \left\{\mathbf{u}_i^* + \left(\mathbf{u}_i^{k+1} - \mathbf{u}_i^*\right)\right\}\Delta t \\
&= \left(\mathbf{r}_i^k + \mathbf{u}_i^*\Delta t\right) + \left(\mathbf{u}_i^{k+1} - \mathbf{u}_i^*\right)\Delta t \\
&= \mathbf{r}_i^* + \left(\mathbf{u}_i^{k+1} - \mathbf{u}_i^*\right)\Delta t
\end{aligned}
\tag{2.53}
$$

Note 2.16: *Advantage of the MPS method in the convection term calculation, and the details of the material derivative $\frac{D}{Dt}$*: As shown in Eqs. (2.47) and (2.52), the MPS method can easily calculate the displacement of particles. These equations are applied to all fluid particles including particles on a free surface without requiring any special procedures. Moreover, free surfaces do not blur.

Additionally, in the MPS method, it is easy to calculate the temporary velocity from Eq. (2.45), while it is more complicated in grid methods. In grid methods, it is necessary to divide the material derivative of a fluid velocity $\frac{D\mathbf{u}}{Dt}$ into the two terms expressed by the right-hand side of the following equation (Note 2.17):

$$
\frac{D\mathbf{u}}{Dt} = \frac{\partial \mathbf{u}}{\partial t} + \mathbf{u} \cdot \nabla \mathbf{u}
\tag{2.54}
$$

where $\frac{\partial \mathbf{u}}{\partial t}$ expresses the time change of the velocity at the position of a calculation point whose position is fixed to a certain space. The second term on the right-hand side of Eq. (2.54) is called a "convective term" or "convective acceleration," and expresses the velocity increment caused by the displacement of the fluid particle.

The convective term is necessary to express the time change of the velocity of a certain fluid particle using the velocity distribution, which is expressed by velocities at fixed calculation points. I would like to explain the convective term by comparing it to water flow in a channel. If we measure the velocity of fluid at a fixed point, such as the inlet of the channel, at every time step, the velocity of a different fluid particle is measured because water flows over the measuring point. Even if $\frac{\partial \mathbf{u}}{\partial t} = 0$ and the current is stationary at a calculation point, which is fixed to space, the velocity of the fluid particle can be accelerated or decelerated at the next time step because of the velocity distribution. For example, if the channel becomes narrower downstream and the fluid has a velocity distribution whose gradient of velocity is positive in the direction of fluid motion, the fluid particle will be accelerated as the fluid particle goes downstream. On the other hand, if the channel becomes wider downstream and the fluid has a velocity distribution whose gradient of velocity is negative in the direction of fluid motion, the fluid particle will decelerate as the fluid particle moves downstream. Of course, additional forces do not actually work on the fluid, although it appears as if additional forces occur at the fixed measuring position because fluid velocity can be accelerated or decelerated by the convection term.

When we measure the acceleration of a fluid particle, a simple way is to compare the particle's velocities at two different times. Specifically, we can calculate the acceleration of a particle by subtracting the velocity of the current time step from that of the next time step and dividing the result of the subtraction by the time difference Δt. That is the way of the MPS method. On the other hand, in grid methods, such as the finite difference method, we must use a velocity distribution whose measuring positions are fixed to a certain space. Therefore, we need to add $\mathbf{u} \cdot \nabla \mathbf{u}$, which is the information of the velocity gradient in the direction of the particle motion, to $\frac{\partial \mathbf{u}}{\partial t}$ to calculate the acceleration of a fluid particle in grid methods. In grid methods, the convection term complicates simulations. In particular, it is complicated to calculate the convection term around free surfaces.

In the MPS method, fluid velocities are not defined at fixed places, but are defined at positions of particles, which flow and move along with fluid flow. Therefore, $\frac{\mathbf{u}_i^{k+1} - \mathbf{u}_i^k}{\Delta t}$ in the MPS method is the material derivative of velocity $\frac{D\mathbf{u}}{Dt}$, which is the left-hand side of Eq. (2.54), while $\frac{\mathbf{u}_i^{k+1} - \mathbf{u}_i^k}{\Delta t}$ expresses $\frac{\partial \mathbf{u}}{\partial t}$ in grid methods. In the MPS method, it is not necessary to divide the acceleration of a particle into two components shown in the right-hand side of Eq. (2.54). Therefore, we need not to calculate the complicated convection term. This is a major advantage of the MPS method.

Note 2.17: *Derivation of Eq. (2.54)*: Vector $\mathbf{u}$, which is the velocity of a fluid particle, is expressed as $\mathbf{u} = \mathbf{u}(x, y, z, t)$, where x, y, and z are coordinates fixed to the space, and t is the time. From the chain rule, we obtain $d\mathbf{u}(x, y, z, t)$, which is the total differential of $\mathbf{u}$, expressed as follows:

$$d\mathbf{u}(x, y, z, t) = \frac{\partial \mathbf{u}}{\partial x} dx + \frac{\partial \mathbf{u}}{\partial y} dy + \frac{\partial \mathbf{u}}{\partial z} dz + \frac{\partial \mathbf{u}}{\partial t} dt \qquad (2.55)$$

The acceleration of a fluid particle is defined as:

$$\mathbf{a} \equiv \frac{D\mathbf{u}}{Dt} = \frac{d\mathbf{u}(x, y, z, t)}{dt} \tag{2.56}$$

so we can express the acceleration of a fluid particle as follows:

$$
\begin{aligned}
\mathbf{a} &= \frac{d\mathbf{u}(x, y, z, t)}{dt} = \frac{\partial \mathbf{u}}{\partial x}\frac{dx}{dt} + \frac{\partial \mathbf{u}}{\partial y}\frac{dy}{dt} + \frac{\partial \mathbf{u}}{\partial z}\frac{dz}{dt} + \frac{\partial \mathbf{u}}{\partial t}\frac{dt}{dt} \\
&= \frac{\partial \mathbf{u}}{\partial x}u_x + \frac{\partial \mathbf{u}}{\partial y}u_y + \frac{\partial \mathbf{u}}{\partial z}u_z + \frac{\partial \mathbf{u}}{\partial t} \\
&= \frac{\partial \mathbf{u}}{\partial t} + \begin{pmatrix} u_x \\ u_y \\ u_z \end{pmatrix} \cdot \begin{pmatrix} \dfrac{\partial \mathbf{u}}{\partial x} \\ \dfrac{\partial \mathbf{u}}{\partial y} \\ \dfrac{\partial \mathbf{u}}{\partial z} \end{pmatrix} = \frac{\partial \mathbf{u}}{\partial t} + \begin{pmatrix} u_x \\ u_y \\ u_z \end{pmatrix} \cdot \begin{pmatrix} \dfrac{\partial}{\partial x} \\ \dfrac{\partial}{\partial y} \\ \dfrac{\partial}{\partial z} \end{pmatrix} \mathbf{u} \\
&= \frac{\partial \mathbf{u}}{\partial t} + \mathbf{u} \cdot \nabla \mathbf{u}
\end{aligned}
\tag{2.57}
$$

From Eqs. (2.56) and (2.57), we can obtain Eq. (2.54).

Note 2.18: *The number of unknown values and the number of equations*: In this note, we add an explanation about why we cannot calculate pressure by the Navier–Stokes equations. On the basis of the relationship between the number of unknown values and the number of equations, the Navier–Stokes equations is discretized as follows:

$$\frac{\mathbf{u}^{k+1} - \mathbf{u}^k}{\Delta t} = -\frac{1}{\rho^0}\langle \nabla P \rangle + \nu \langle \nabla^2 \mathbf{u}^k \rangle + \mathbf{g} \tag{2.58}$$

The vector, $\mathbf{u}^{k+1}$, which is the velocity vector at time step $k + 1$, is unknown because it is a future value. Pressure P is also an unknown value. The time step number of pressure is not written in Eq. (2.44) because the time step number of pressure depends on the simulation algorithm, although the time step number of pressure is $k + 1$ in the case where the semi-implicit algorithm is applied. The time step number is also not written for the gravity term $\mathbf{g}$ because the gravity term is constant. In three-dimensional simulations, $\mathbf{u}^{k+1}$ has three components: x-, y-, and z-directional components. Therefore, the number of unknowns is four including pressure. There are only three conditional equations, Eqs. (2.9), (2.10), and (2.11), which are the Navier—Stokes equations in component expression. By comparing the number of unknown values and the number of conditional equations, we find that the number of unknown values is larger than that of conditional equations by one, and it is difficult to determine the unknown values. Therefore, we need to add the equation of continuity (Eq. 2.7) or the equation of state (see Eq. 3.38 in Section 3.2 and Program 2.21) as the additional conditional equation. Thus, we calculate the pressure by solving the pressure Poisson equation (Eq. 2.49), which is derived using the equation of continuity as a conditional equation. The details about the derivation of the pressure Poisson equation are explained in Section 2.2.5.4.

2.2.5.4 Derivation of Pressure Poison Equation of the MPS Method

As mentioned earlier, we must add a conditional equation because the number of conditional equations is fewer than that of unknown values by one. Thus, we use the equation of continuity (Eq. 2.7) as the additional conditional equation. If the fluid density is assumed to be constant, the material derivative of fluid density is expressed as follows:

$$\frac{D\rho}{Dt} = 0 \tag{2.59}$$

In this case, the equation of continuity (Eq. 2.7) is expressed as follows:

$$\nabla \cdot \mathbf{u} = 0 \tag{2.60}$$

We derive an equation, which determines fluid pressure, by combining Eq. (2.46) and the Navier—Stokes equation. The Navier—Stokes equations are expressed as follows:

$$\frac{D\mathbf{u}_i}{Dt} = -\frac{1}{\rho^0}\langle \nabla P \rangle_i^{k+1} + \nu \langle \nabla^2 \mathbf{u} \rangle_i^k + \mathbf{g} \tag{2.61}$$

where symbols $\langle \rangle_i$ are used for pressure gradient and the Laplacian of fluid velocity because they are approximated by the models in the MPS method at the position of the ith particle. As explained in Sections 2.2.5.2 and 2.2.5.3, we calculate Eq. (2.61) on the basis of the fractional step method, which integrates Eq. (2.61) in two steps as follows:

$$\frac{\mathbf{u}_i^* - \mathbf{u}_i^k}{\Delta t} = + \nu \langle \nabla^2 \mathbf{u} \rangle_i^k + \mathbf{g} \tag{2.62}$$

$$\frac{\mathbf{u}_i^{k+1} - \mathbf{u}_i^*}{\Delta t} = -\frac{1}{\rho^0} \langle \nabla P \rangle_i^{k+1} \tag{2.63}$$

In order to use Eq. (2.60) as a conditional equation, we calculate the divergence of both side of Eq. (2.63), and obtain the following equation:

$$\frac{\langle \nabla \cdot \mathbf{u} \rangle_i^{k+1} - \langle \nabla \cdot \mathbf{u}^* \rangle_i^*}{\Delta t} = -\frac{1}{\rho^0} \langle \nabla \cdot \nabla P^{k+1} \rangle_i \tag{2.64}$$

To satisfy the incompressible condition and the equation of continuity at the time step of $k+1$, Eq. (2.60) should be satisfied. Therefore, Eq. (2.50) is expressed as follows:

$$\frac{0 - \langle \nabla \cdot \mathbf{u} \rangle_i^*}{\Delta t} = -\frac{1}{\rho^0} \langle \nabla^2 P \rangle_i^{k+1} \tag{2.65}$$

where we used the incompressible condition $\langle \nabla \cdot \mathbf{u} \rangle_i^{k+1} = 0$ and the Laplacian ∇^2, which is equal to $\nabla \cdot \nabla$. We find that the unknown vector, $\mathbf{u}^{k+1}$, has disappeared. By rearranging Eq. (2.65), we obtain the following pressure Poisson equation:

$$\langle \nabla^2 P \rangle_i^{k+1} = \rho^0 \frac{\langle \nabla \cdot \mathbf{u} \rangle_i^*}{\Delta t} \tag{2.66}$$

In grid-based methods, we usually calculate fluid pressure by solving Eq. (2.66). However, in the case of the MPS method, fluid volume might not be conserved well and may become lower than the initial value because the source term, which is the right-hand side of Eq. (2.66), is expressed by the divergence of velocity. Thus the calculation error can be accumulated as time passes. So, we transform the right-hand side of Eq. (2.66). Eq. (2.7), which is the equation of continuity, is discretized as follows:

$$\frac{\rho_i^* - \rho^0}{\Delta t} + \rho^0 \langle \nabla \cdot \mathbf{u} \rangle_i^* = 0 \tag{2.67}$$

where ρ_i^* is the fluid density, which is calculated in the particle distribution of $\mathbf{r}^*$, and corresponds to the state in Fig. 2.10B. We approximate $D\rho/Dt$, which is the material derivative of fluid density, by the forward Euler method. The above equation is transformed as follows:

$$\langle \nabla \cdot \mathbf{u} \rangle_i^* = -\frac{1}{\Delta t} \frac{\rho_i^* - \rho^0}{\rho^0} \cong -\frac{1}{\Delta t} \frac{n_i^* - n^0}{n^0} \tag{2.68}$$

where we use the approximation of Eq. (2.20) to derive Eq. (2.68). The divergence of the velocity, which is the left-hand side, is approximated by the deviation of the particle number density during Δt as shown in the right-hand side of Eq. (2.68). By substituting the above equation for the right-hand side of Eq. (2.66), we obtain the following equation:

$$
\begin{aligned}
\left\langle \nabla^2 P \right\rangle_i^{k+1} &= \rho^0 \frac{1}{\Delta t} \left(-\frac{1}{\Delta t} \frac{n_i^* - n^0}{n^0} \right) \\
&= -\frac{\rho^0}{(\Delta t)^2} \left(\frac{n_i^* - n^0}{n^0} \right)
\end{aligned}
\tag{2.69}
$$

The above equation is the pressure Poisson equation of the MPS method (Eq. 2.49). The source term, which is the right-hand side of Eq. (2.69), is expressed by the particle number density. Therefore, we can detect the accumulation of the density error. As a result, volume conservation is improved. This is because a compression state, where distance between particles is much shorter than the initial value, is detectable using the weight function of the particle number density. If the source term is expressed by the velocity divergence as shown in Eq. (2.66), the accumulation of density error cannot be detected. There are other ways to express the source term of the pressure Poisson equation. For example, some studies (Ikeda, 1999; Tanaka and Masunaga, 2010) use a mixed source term, which is expressed by both the velocity divergence and the change rate of the particle number density. These studies reported that the mixed source term is able to restrain the pressure oscillation effectively.

2.2.5.5 How to Calculate the Pressure Poisson Equation

Let us study how to calculate the pressure Poisson equation (Eq. 2.69).By dividing both sides of Eq. (2.38) by ρ^0, we obtain the following equation:

$$
\frac{1}{\rho^0} \left\langle \nabla^2 P \right\rangle_i^{k+1} = -\frac{1}{(\Delta t)^2} \frac{n_i^* - n^0}{n^0}
\tag{2.70}
$$

Note that ρ^0 was moved from the right-hand side to the left-hand side. Then, we apply the Laplacian model of the MPS method, which is Eq. (2.33), to discretize the Laplacian operator of the left-hand side of Eq. (2.70) as follows:

$$
\frac{1}{\rho^0} \frac{2d}{\lambda^0 n^0} \sum_{j \neq i} \left(P_j^{k+1} - P_i^{k+1} \right) w \left(\left| r_j^* - r_i^* \right|, r_e \right) = -\frac{1}{(\Delta t)^2} \frac{n_i^* - n^0}{n^0}
\tag{2.71}
$$

We find that the partial differential operator ∇^2 is changed into basic operators, which are executable on computers. In Eq. (2.71), the unknown parameters are the pressures P_i^{k+1} and P_j^{k+1}, which are future values. The other values are constant, or are obtained by easy calculations.

Let us simplify the expression of Eq. (2.71) as follows:

$$a_{i1}P_1^{k+1} + a_{i2}P_2^{k+1} + \cdots + a_{ii}P_i^{k+1} + \cdots + a_{iN}P_N^{k+1} = b_i \tag{2.72}$$

where

$$a_{ij} = \begin{cases} \dfrac{1}{\rho^0} \dfrac{2d}{\lambda^0 n^0} w\left(\left|\mathbf{r}_j^* - \mathbf{r}_i^*\right|, r_e\right) & (j \neq i) \\[4mm] -\dfrac{1}{\rho^0} \dfrac{2d}{\lambda^0 n^0} \displaystyle\sum_{j' \neq i} w\left(\left|\mathbf{r}_{j'}^* - \mathbf{r}_i^*\right|, r_e\right) & (j = i) \end{cases} \tag{2.73}$$

$$b_i = -\frac{1}{(\Delta t)^2} \frac{n_i^* - n^0}{n^0} . \tag{2.74}$$

The parameter N is the total number of particles. Each particle has a relationship, which is expressed in the same manner as Eq. (2.72). Therefore, the entire relationship among particles is expressed by the following simultaneous linear equations, whose unknown values are particle pressures at time step $k + 1$.

$$\begin{aligned} a_{11}P_1^{k+1} + a_{12}P_2^{k+1} + \cdots + a_{1i}P_i^{k+1} \cdots + a_{1N}P_N^{k+1} &= b_1 \\ a_{21}P_1^{k+1} + a_{22}P_2^{k+1} + \cdots + a_{2i}P_i^{k+1} \cdots + a_{2N}P_N^{k+1} &= b_2 \\ &\vdots \\ a_{i1}P_1^{k+1} + a_{i2}P_2^{k+1} + \cdots + a_{ii}P_i^{k+1} \cdots + a_{iN}P_N^{k+1} &= b_i \\ &\vdots \\ a_{N1}P_1^{k+1} + a_{N2}P_2^{k+1} + \cdots + a_{Ni}P_i^{k+1} \cdots + a_{NN}P_N^{k+1} &= b_N \end{aligned} \tag{2.75}$$

Using a matrix expression, we can write Eq. (2.75) compactly as follows:

$$\begin{pmatrix} a_{11} & a_{12} & \cdots & a_{1j} & \cdots & a_{1N} \\ a_{21} & a_{22} & \cdots & a_{2j} & \cdots & a_{2N} \\ \vdots & \vdots & & \vdots & & \\ a_{i1} & a_{i2} & \cdots & a_{ij} & \cdots & a_{iN} \\ \vdots & \vdots & & \vdots & & \\ a_{N1} & a_{N2} & \cdots & a_{2j} & \cdots & a_{NN} \end{pmatrix} \begin{pmatrix} P_1^{k+1} \\ P_2^{k+1} \\ \vdots \\ P_i^{k+1} \\ \vdots \\ P_N^{k+1} \end{pmatrix} = \begin{pmatrix} b_1 \\ b_2 \\ \vdots \\ b_i \\ \vdots \\ b_N \end{pmatrix} \tag{2.76}$$

Then we can simply write Eq. (2.72) as follows:

$$\mathbf{Ax} = \mathbf{b} \qquad (2.77)$$

where

$$\mathbf{A} = \begin{pmatrix} a_{11} & a_{12} & \cdots & a_{1j} & \cdots & a_{1N} \\ a_{21} & a_{22} & \cdots & a_{2j} & \cdots & a_{2N} \\ \vdots & \vdots & & \vdots & & \\ a_{i1} & a_{i2} & \cdots & a_{ij} & \cdots & a_{iN} \\ \vdots & \vdots & & \vdots & & \\ a_{N1} & a_{N2} & \cdots & a_{2j} & \cdots & a_{NN} \end{pmatrix} \qquad (2.78)$$

$$\mathbf{x} = \begin{pmatrix} P_1^{\,k+1} \\ P_2^{\,k+1} \\ \vdots \\ P_i^{\,k+1} \\ \vdots \\ P_N^{\,k+1} \end{pmatrix} \qquad (2.79)$$

$$\mathbf{b} = \begin{pmatrix} b_1 \\ b_2 \\ \vdots \\ b_i \\ \vdots \\ b_N \end{pmatrix} \qquad (2.80)$$

After setting $\mathbf{A}$ and $\mathbf{b}$ in Eqs. (2.73) and (2.74), we can solve the simultaneous linear equations by an arbitrary solver, such as the conjugate gradient method, Gaussian elimination, Jacobi method, SOR, etc., and obtain the solution for $\mathbf{x}$, which is pressure P^{k+1} (see Eq. 2.79).

Note 2.19: *The features of the coefficient matrix* $\mathbf{A}$ *of the pressure Poisson equation in the MPS method*: The coefficient matrix $\mathbf{A}$ of the pressure Poisson equation is a sparse matrix, in which most elements are zero, because the weight function used in the Laplacian model of the MPS method becomes zero if the distance between particle i and particle j is longer than or equal to the effective radius r_e. Moreover, $\mathbf{A}$ is symmetric if we apply the Laplacian model of the MPS method and the standard boundary condition because the Laplacian model is symmetric between particle *i* and particle *j*. Therefore, we can use a fast solver, such as the conjugate gradient method, for solving the simultaneous linear equations.

2.2.5.6 The Boundary Condition of Pressure

We need to decide on boundary conditions for solving the pressure Poisson equation. We usually use the following two types of boundary conditions. One is the Dirichlet boundary condition, which directly fixes some parameters to certain values. The Dirichlet boundary condition works as the standard value of the pressure distribution in the MPS method. We express particle pressures as relative values whose standard value is 0 Pa on free surfaces. That is, we assign 0 Pa to particles on free surfaces (free-surface particles) as a Dirichlet boundary condition. Although we can assign the atmospheric pressure, which is about 1013 hPa, instead of 0 Pa, we usually use 0 Pa because it is easy to compare the simulated pressure with the gauge pressure of an experiment. Gauge pressure used in experiments is the pressure expressed as the difference from the atmospheric pressure. In the Navier–Stokes equation, the pressure is used to calculate the pressure gradient. The pressure gradient is a relative value, which does not depend on a standard value. Therefore, we can use 0 Pa for the pressure of free-surface particles. There are some techniques for judging whether each particle is a free-surface particle or not. For details, see Section 4.3.1. The most traditional technique uses the particle number density for judging free-surface particles (Koshizuka and Oka, 1996). If the particle number density of a particle satisfies the following condition, the particle is regarded as a free-surface particle and is assigned 0 Pa.

$$n_i < \beta n^0 \tag{2.81}$$

where β is an empirically-determined scalar parameter. In many cases, 0.97 is used for β.

The second type of boundary condition is the Neumann boundary condition. This is a boundary condition expressed by first derivatives. Neumann boundary conditions are used for wall boundaries in the pressure calculation. In the MPS method, a Neumann boundary condition is that a zero pressure gradient is assigned to wall boundaries. To approximately express the Neumann boundary condition, two layers of dummy wall particles, which do not have pressure parameters, are located behind a single or double layer of wall particles, which have pressure parameters.

Using the Neumann boundary condition, we can avoid penetration of fluid particles through walls. The details are explained in Section 2.5.2.

2.2.5.7 The Boundary Condition of Velocity

The no-slip boundary condition of a stationary wall is expressed by setting the velocities of wall particles to 0 m/s. The sample programs in this book use this technique. To express more accurate no-slip boundary conditions, some researchers assign a mirror velocity distribution to wall particles in the same manner as the finite difference method. For details, see Section 4.2.2.

The slip boundary condition is expressed by omitting neighboring wall particles in the viscosity calculation of Eq. (2.46), which is the viscosity calculation of a fluid particle. That is, if neighboring particle j is a wall particle, we do not calculate the particle interaction between fluid particle i and wall particle j because viscosity force does not exert an influence between the fluid and a slip wall boundary.

2.3 OUTLINE OF SIMULATION PROGRAMS

This section explains the source codes of the sample programs of the MPS method and the explicit MPS method. You may download the source codes from the following website:

URL: https://mpsbook2018.wixsite.com/mysite

ID: mpsuser

Password: e7A5B3Gm

File names of source codes:

1. mps.c—Source code of the MPS method, in which pressure is implicitly calculated.
2. emps.c—Source code of the explicit MPS method, in which pressure is explicitly calculated.

The meanings of functions, defined values, and variables in the programs are explained in Tables 2.1, 2.2, and 2.3, respectively.

2.3.1 Contents of Program

The sample program "mps.c" can simulate incompressible flows on the basis of the MPS method. Fig. 2.1 is the simulation result of the sample program. Another sample program "emps.c" can simulate free-surface flows on the basis of the explicit MPS method. Most parts of both programs are the same. The parts that differ are the pressure calculation (Program

Table 2.1 List of functions

Function name	Argument	Return type	Contents of processing
calculateGravity()	void	void	Set the array of particle acceleration to the acceleration of gravity.
calculateNumberDensity()	void	void	Calculate the particle number density.
calculateNZeroAndLambda()	void	void	Calculate n^0 and λ^0
calculatePressure()	void	void	Calculate pressure.
calculatePressureGradient()	void	void	Calculate the acceleration due to the pressure gradient term.
calculatePressureGradient_forExplicitMPS()	void	void	Calculate the acceleration due to the pressure gradient term (a function for the explicit MPS method)
calculatePressuret_forExplicitMPS()	void	void	Explicitly calculate fluid pressure on the basis of the equation of state (a function for the explicit MPS method)
calculateViscosity()	void	void	Calculate the acceleration due to the viscosity term.
checkBoundaryCondition()	void	void	Check whether Dirichlet boundary conditions are appropriately given to the pressure Poisson equation.
collision()	void	void	Calculate collisions between particles as an exceptional processing.
exceptionalProcessing ForBoundaryCondition()	void	void	Carry out an exceptional processing for particle group which do not have any Dirichlet boundary conditions.
increaseDiagonalTerm()	void	void	Increase some of the diagonal terms of the coefficient matrix of the simultaneous linear equations as an exceptional processing for particle groups which do not have any Dirichlet boundary conditions.
initializeParticlePosition AndVelocity_for2dim()	void	void	Initialize positions and velocities of particles in two-dimensional simulation.
initializeParticlePosition AndVelocity_for3dim()	void	void	Initialize positions and velocities of particles in three-dimensional simulation.
main()	int, char**	int	Main function (the function which is executed first in the program).
mainLoopOfSimulation()	void	void	Loop function in which procedures for a time step are repeatedly executed until the simulation time reaches the finish time.
moveParticle()	void	void	Update particle velocities and move particles on the basis of the particle accelerations which consist of the gravity and viscosity terms.

(Continued)

Table 2.1 (Continued)

Function name	Argument	Return type	Contents of processing
moveParticleUsingPressureGradient()	void	void	Correct particle velocities and correct particle positions on the basis of the pressure gradients.
removeNegativePressure()	void	void	Set negative pressures to 0 Pa.
setBoundaryCondition()	void	void	Give Dirichlet boundary conditions for solving the pressure Poisson equation.
setConstantParameter()	void	void	Set constant parameters, such as effective radii.
setMatrix()	void	void	Set coefficient matrix of the simultaneous linear equation.
setMinimumPressure()	void	void	Set the lowest pressure among neighboring particles and ith particle.
setSourceTerm()	void	void	Set source term, which is the right-hand side of pressure Poisson equation.
solveSimultaniousEquations ByGaussianElimination()	void	void	Solve the simultaneous linear equation about pressure by Gaussian elimination.
weight()	double, double	double	Calculate the weight of a neighboring particle by the weight function.
writeData_inProfFormat()	void	void	Output the simulation result of the current time step into a file in PROF format, which is an original file format made by the authors. In *.prof files (output_0000.prof, output_0001.prof, .), simulation time is written in the first line. The number of particles is written in the second line. From the third line onwards, the particle information is written in a line. In each line, the program outputs the x-, y-, and z-coordinates, x-, y-, and z-components of velocity, pressure, and the particle number density of a particle.
writeData_inVtuFormat()	void	void	Output the simulation result of the current time step into a file (particle_0000.vtu, particle_0001.vtu ...) in VTU format, which is a ParaView file format.

Symbolic constants	Replacement-text	Meaning	Unit
ARRAY_SIZE	5000	Size of an array (upper limit of the number of particles).	None
COEFFICIENT_OF_RESTITUTION	0.2	Coefficient of restitution.	None
COMPRESSIBILITY	$(-0.45\text{E} - 9)$	Compressibility of fluid.	1/Pa
DIM	2	Number of space dimensions (in two-dimensional simulation, DIM is 2. In three-dimensional simulation, DIM is 3).	None
DIRICHLET_BOUNDARY_IS_CHECKED	2	Constant of flag for checking whether a Dirichlet boundary condition is appropriately given.	None
DIRICHLET_BOUNDARY_IS_CONNECTED	1	Constant of flag for checking whether a Dirichlet boundary condition is appropriately given.	None
DIRICHLET_BOUNDARY_IS_NOT_CONNECTED	0	Constant of flag for checking whether a Dirichlet boundary condition is appropriately given.	None
DT	0.001	Time increment Δt.	s
DUMMY_WALL	2	Type number of dummy particle.	None
EPS	(0.01*PARTICLE_DISTANCE)	Small value for arranging particle on boundary of a region.	m
FINISH_TIME	2	Finish time of simulation.	s
FLUID	0	Type number of fluid particle.	None
FLUID_DENSITY	1000	Fluid density ρ^0.	kg/m^3
GHOST	(-1)	Type number of ghost particle.	None
GHOST_OR_DUMMY	(-1)	Constant which is used for judging whether the pressure calculation should be skipped or not. The pressure calculations of ghost or dummy wall particles are skipped because the pressures are not used.	None
GRAVITY_X	0	x-component of the acceleration of the gravity.	m/s^2
GRAVITY_Y	(-9.8)	y-component of the acceleration of the gravity	m/s^2

(Continued)

Table 2.2 (Continued)

Symbolic constants	Replacement-text	Meaning	Unit
GRAVITY_Z	0	z-component of the acceleration of the gravity.	m/s^2
INNER_PARTICLE	0	Constant which is used for judging whether the particle pressure should be fixed or updated. Pressure of a free-surface particle is fixed to 0 Pa.	None
KINEMATIC_VISCOSITY	(1.0E − 6)	Kinematic viscosity coefficient.	m^2/s
OFF	0	Constant for expressing a flag is false.	None
ON	1	Constant for expressing a flag is true.	None
OUTPUT_INTERVAL	20	Interval for outputting the simulation results in files.	Time steps
PARTICLE_DISTANCE	0.025	Initial distance between particles l_0.	m
RADIUS_FOR_GRADIENT	(2.1*PARTICLE_DISTANCE)	Radius of the interaction zone for the gradient model.	m
RADIUS_FOR_LAPLACIAN	(3.1*PARTICLE_DISTANCE)	Radius of the interaction zone for the Laplacian model	m
RADIUS_FOR_NUMBER_DENSITY	(2.1*PARTICLE_DISTANCE)	Radius of the interaction zone for particle number density calculation.	m
RATIO_OF_COLLISION_DISTANCE	0.5	Nondimensional threshold distance for collision calculation.	None
RELAXATION_COEFFICIENT_FOR_PRESSURE	0.2	Coefficient multiplied to the source term of the pressure Poisson equation for restraining the pressure oscillation.	None
SPEED_OF_SOUND	10	Speed of sound (a symbolic constant for the explicit MPS method).	m/s
SURFACE_PARTICLE	1	Constant for giving a Dirichlet boundary condition of the pressure Poisson equation to a free-surface particle.	None
THRESHOLD_RATIO_OF_PARTICLE_NUMBER_DENSITY	0.97	Constant used for judging free-surface particles.	None
WALL	1	Type number of wall particle.	None

Table 2.3 List of variables

Variable name	Type	Meaning of the variable	Unit
a	double	Coefficient used for calculating models of partial differential derivatives	
absoluteValueOfVelocity	double	Absolute value of velocity	m/s
AccelerationX[ARRAY_SIZE]	double	Acceleration of particles in x-direction	m/s^2
AccelerationY[ARRAY_SIZE]	double	Acceleration of particles in y-direction	m/s^2
AccelerationZ[ARRAY_SIZE]	double	Acceleration of particles in z-direction	m/s^2
alpha	double	Compressibility of fluid	1/Pa
argc	int	Argument count (the number of arguments for main function)	
argv	char**	Argument vector (list of arguments for main function)	
BoundaryCondition[ARRAY_SIZE]	int	Flag for giving Dirichlet boundary conditions	
c	double	Speed of sound (a parameter for the explicit MPS method)	m/s
c2	double	Square of the speed of sound (a parameter for the explicit MPS method)	m^2/s^2
coefficientIJ	double	Element of the coefficient matrix of the simultaneous linear equations	
CoefficientMatrix[ARRAY_SIZE* ARRAY_SIZE]	double	Coefficient matrix of the simultaneous linear equations	
collisionDistance	double	Threshold distance for collisions (if two particles are getting close, and the distance between particles is shorter than the collisionDistance, a collision, which is an exceptional procedure, occurs between them)	m
collisionDistance2	double	Square of collisionDistance	m^2
count	int	Counter of particles	
distance	double	Distance between particles	m
distance2	double	Square of distance between particles	m^2
fileName[1024]	char	File name of simulation results	
FileNumber;	int	Counter of output files	
FlagForCheckingBoundaryCondition [ARRAY_SIZE];	int	Flag for checking whether a Dirichlet boundary condition is given to the fluid group	
flagOfParticleGeneration;	int	Flag for particle generation	

(Continued)

Table 2.3 (Continued)

Variable name	Type	Meaning of the variable	Unit
FluidDensity;	double	Distance between particles	kg*m^3
forceDT;	double	Impulse of collision (forceDt = force * deltaTime)	
fp;	FILE*	File pointer	
Fp;	FILE*	File pointer	
gradientX,	double	Gradient in x-direction	Pa/m
gradientY,	double	Gradient in y-direction	Pa/m
gradientZ;	double	Gradient in z-direction	Pa/m
i	int	Identification number of particle or index used for repeated procedures	
iTimeStep	int	Counter of time steps (identification number of time step)	
iX,	int	Index of column in x-direction	
iY;	int	Index of column in y-direction	
iZ;	int	Index of column in z-direction	
iZ_end;	int	Upper limit of a parameter, iZ	
iZ_start,	int	Lower limit of a parameter, iZ	
j	int	Identification number of neighboring particle, or index used for repeated procedures	
Lambda;	double	λ^0 of Laplacian model	m^2
mi;	double	Mass of ith particle	kg
MinimumPressure[ARRAY_SIZE];	double	Lowest pressure in the interaction zone of each particle	Pa
mj	double	Mass of a neighboring particle j	kg
n	int	Number of columns or row in a matrix	
n0;	double	Standard value of the particle number density, n0 (or the standard value of the total weight)	
N0_forGradient;	double	n^0 for gradient model	
N0_forLaplacian;	double	n^0 for Laplacian model	
N0_forParticleNumberDensity;	double	Standard value of the particle number density, n0	

Variable	Type	Description	Units
NumberOfParticles	int	Total number of particles	
nX,	int	Number of particle layers in x-directions	
nY,	int	Number of particle layers in y-directions	
nZ;	int	Number of particle layers in z-directions	
ParticleNumberDensity[ARRAY_SIZE];	double	Particle number density	
ParticleType[ARRAY_SIZE];	int	Particle type	
pij;	double	Pressure difference between particle j and particle i ($Pij = Pj - Pi$)	Pa
PositionX[ARRAY_SIZE];	double	X-coordinates of particles	m
PositionY[ARRAY_SIZE];	double	Y-coordinates of particles	m
PositionZ[ARRAY_SIZE];	double	Z-coordinates of particles	m
Pressure[ARRAY_SIZE];	double	Particle pressure	Pa
Radius_forGradient,	double	Effective radius of interaction zone for gradient model	m
Radius_forLaplacian,	double	Effective radius of interaction zone for Laplacian model	m
Radius_forNumberDensity	double	Effective radius of interaction zone for particle number density	m
Radius2_forGradient;	double	Square of effective radius of interaction zone for gradient model	m^2
Radius2_forLaplacian;	double	Square of effective radius of interaction zone for Laplacian model	m^2
Radius2_forNumberDensity	double	Square of effective radius of interaction zone for particle number density	m^2
rho	double	Fluid density (a parameter for the explicit MPS method)	kg/m^3
SourceTerm[ARRAY_SIZE];	double	Source term of the pressure Poisson equation	
sumOfTerms;	double	Sum of terms	
Time;	double	Simulation time	s
velocity_ix,	double	x-component of velocity vector of ith particle	m/s
velocity_iy,	double	y-component of velocity vector of ith particle	m/s
velocity_iz;	double	z-component of velocity vector of ith particle	m/s
VelocityX[ARRAY_SIZE];	double	Velocity of particle in x-direction	m/s
VelocityY[ARRAY_SIZE];	double	Velocity of particle in y-direction	m/s

(Continued)

Table 2.3 (Continued)

Variable name	Type	Meaning of the variable	Unit
VelocityZ[ARRAY_SIZE];	double	Velocity of particle in z-direction	m/s
viscosityTermX,	double	x-component of acceleration vector due to the viscosity term	m/s^2
viscosityTermY,	double	y-component of acceleration vector due to the viscosity term	m/s^2
viscosityTermZ;	double	z-component of acceleration vector due to the viscosity term	m/s^2
w	double	Weight of a neighboring particle j	
weightIJ;	double	Weight of a neighboring particle j	
x,	double	x-coordinate	m
xi	double	x-coordinate of ith particle	m
xij	double	x-component of a relative coordinate ($xij = xj - xi$)	m
xj	double	x-coordinate of a neighboring particle j	m
y	double	y-coordinate	m
yi	double	y-coordinate of ith particle	m
yij	double	y-component of a relative coordinate ($yij = yj - yi$)	m
yj	double	y-coordinate of a neighboring particle j	m
z	double	z-coordinate	m
zi	double	z-coordinate of ith particle	m
zij	double	z-component of a relative coordinate ($zij = zj - zi$)	m
zj	double	z-coordinate of a neighboring particle j	m

2.21) and the pressure gradient calculation (Program 2.22). For details of the explicit MPS method, see Sections 2.3.4.17, 2.3.4.18, and 3.2.

A dam break, in which a water column collapses by gravity, is simulated in two or three dimensions using a sample program. We can simulate various fluid phenomena other than a dam break by changing the initial condition of a program. Viscosity, gravity, and pressure are taken into account. The semi-implicit algorithm explained in Section 2.2.5 is applied to mps.c. Input files are unnecessary for both sample programs because the coordinates of particles and the initial fluid velocities are determined by the programs. The programs are written in the C language, and are developed for understanding the basic theory of the MPS method. To simplify the programs, techniques for shortening computational time are not introduced. We can add these techniques to the source codes after understanding the simple source codes. For example, we can add a fast solver such as the conjugate gradient method, for solving the simultaneous linear equations in mps.c, if necessary.

In the following pages, only the important parts of the source code are written. The omitted parts are denoted by characters, such as "...". You may download the complete source codes without omission from the above website. You might want to print and read the source codes. If you are not familiar with particle simulations, you should read "emps.c" before reading "mps.c" because "emps.c" is shorter and easier to understand than "mps.c." The program "emps.c" is about 700 lines in length, while "mps.c" is about 850 lines.

Note 2.20: *Techniques for shortening the computation time*: As mentioned earlier, to simplify the sample program, we did not add functions for shortening computation time, such as a neighboring particles list and a fast solver for the simultaneous linear equations.

The sample programs calculate particle interactions between each particle and all other particles. Therefore, the calculation amount for particle approximation (e.g., Eqs. 2.17, 2.24, and 2.23) are in proportion to N^2, where N is the number of particles. To reduce the calculation amount, a neighboring particles list is very effective. From the neighboring particle list, we can restrict interacting particles to only neighboring particles, and reduce the calculation amount to the order of N. Constructing the neighboring particle list is also a time-consuming procedure. Typically, we adopt a domain decomposition technique to reduce the computational cost for creating the neighboring particle list. Using the domain decomposition, we can reduce the computational amount from the N^2 order to the N order.

The coefficient matrix $\mathbf{A}$ of the simultaneous linear equations in Eq. (2.78) is a sparse matrix, which contains many zero elements. Although we usually do not save the zero elements in memory to reduce the amount of data, in the sample program, we save all elements including zero elements to simplify the program. This program uses Gaussian elimination, which is one of the basic methods for solving simultaneous linear equations. Therefore, the computational amount for solving the simultaneous equation is in proportion to N^3. If we apply the conjugate gradient method instead of Gaussian elimination, we can reduce the computational amount for solving the simultaneous linear equation of pressure to the $N^{1.5}$ order of magnitude.   .

2.3.2 How to Compile and Execute the Sample Programs

Any compilers of the C language can convert the source code to a binary code, which can run on your computer. (Compiling means to convert the source code of a program into a form which can be executed on a computer. The compiler is software used for compiling.) The fundamental functions of the C language are used in the program. In this section we study how to compile and execute the sample programs of the MPS method.

1. In the case of the Windows environment

   We can use free compilers, such as Microsoft Visual Studio Community, Visual Studio Express, gcc (GNU Compiler Collection) of Cygwin, and gcc of MinGW.

   1.1. If you use Microsoft Visual Studio Community or Express, first create an empty project, then move a sample code to the created project folder. Then, add the source code of the sample program to the created project by dragging the icon of mps.c on Windows Explorer and release the source code over the icon of the source file folder on the solution explorer of the Visual Studio. Then, build the solution (compile the source code), and start without debugging (execute the compiled program). For details, refer to the manual of a compiler.

1.2. If you use gcc of Cygwin or gcc of MinGW, you can compile the downloaded source code (mps.c) and make an executable file (a.exe) by executing the following command (You need to place mps.c or emps.c in the current directory to compile the source code. If you would like to enhance the speed of the simulation program, it would be better to add the compile option "-O3" for optimization as follows: gcc -O3 mps.c −lm, where "-O3" is minus, capital letter "O", and three.) on a terminal:

```
gcc mps.c -lm
```

where "-" is minus, "lm" are the lowercase letters of "LM." The "-lm" option is necessary because math.h, which is a mathematical library, is used in the simulation program. Then, we can execute the program by the following command:./a.exe

In the same manner, we can compile and execute "emps.c" by the following commands:

```
gcc emps.c -lm
./a.exe
```

2. In the cases of UNIX environments (Linux, Mac OS X, etc.)

We can use a free compiler, such as gcc (GNU Compiler Collection). We need to install gcc to use it. For example, if you use Mac OS X, you need to install Xcode and start Xcode once. After gcc is installed, we can compile the downloaded source code (mps.c) and create an executable file (a.out) by executing the following command on a terminal:

```
gcc mps.c -lm
```

where "-" is minus, "lm" is the lowercase for "LM." Again, "-lm" is necessary because math.h is used in the simulation program. Then, we can execute the program by the following command:

```
./a.out
```

In the same manner, we can compile and execute "emps.c" by the following commands:

gcc emps.c -lm
```
./a.out
```

2.3.3 How to Visualize the Simulation Result

We use ParaView, which is free-software, for visualizing simulation results. After starting ParaView, open the simulation results of the MPS method. The file names of the simulation results are particle_0000.vtu to particle_0100.vtu. We can select all the simulation results at once by choosing the icon "particle_. vtu" in the "Open File" dialog (Note 2.20).

Then, click button "Apply." There are various ways for drawing particles. The authors usually use "Point Gaussian" for the "Representation." The size of particles can be adjusted by the parameter "Gaussian Radius." We can change the particle colors by the "Coloring" item. If you change the coloring item from "Solid Color" to "ParticleType", particles are colored by particle type. If you choose "Pressure" for "Coloring," particles are colored by pressure values. If you choose "Velocity" for "Coloring," particles are colored by the absolute values of the particle velocities. If you choose "ParticleNumberDensity" for "Coloring," particles are colored by the particle number densities.

We can watch the simulation results as an animation by clicking the play button, which is a triangle icon. For details about ParaView, see the website (Note 2.21).

Note 2.21: *ParaView*: ParaView is a very fast visualization software and is executable on Windows, Linux, and Mac OS. We can download ParaView for free from the website below. Note that it would be better to use only alphabet and number characters for file names, folder names, and user names because some problems might occur if special characters are used: https://www.paraview.org/.

2.3.4 Functions of the Program
2.3.4.1 Libraries and Declarations
In lines 1−2 of Program 2.5, there are "#include" statements. These statements read required libraries, which are collected general-purpose programs. The libraries are "stdio.h" for input-and-output functions and "math.h" for mathematical functions. We do not need to install any special software or libraries because the two libraries are standard libraries of the C language.

Note 2.22: *#define*: In lines 4−9, there are "#define" statements, which define values or character strings. For example, in line 4 we define the character string named "DIM" as 2, where DIM is the number of spatial dimensions. Therefore, in this case, the simulation is two-dimensional.

```
001   #include <stdio.h>
002   #include <math.h>
003
004   #define DIM                  2
005   #define PARTICLE_DISTANCE    0.025
006   #define DT                   0.001
007   #define OUTPUT_INTERVAL      20
008   #define ARRAY_SIZE           5000
009   # define FINISH_TIME         2.0
010   ...
```

Program 2.5

In the case of three-dimensional simulation, we define DIM as 3. The definition statement allows us to write "DIM" instead of writing a concrete value, such as 2 or 3, and makes the program easy to understand. In the sample program, we calculate three coordinate components, x, y, and z in both two- and three-dimensional simulations to simplify the program, although actually we do not need to calculate z components in two-dimensional simulations.

In two-dimensional simulation, velocity components in the z direction and z coordinates of particle positions are set to zero because the acceleration components in the z-direction are zero.

Note 2.23: *Units of variables*: In this program, the SI unit system (International System of Units) is used. Therefore, length, mass, and time are expressed by "meter." "kilogram," and "second." respectively. For example, line 5 defines "PARTICLE_DISTANCE," which is the initial distance between particles l_0, as 0.025 m.

"DT" in line 6 is the time increment Δt, which represents the time difference between time steps. In the sample program, Δt is set to 0.001 s. "OUTPUT_INTERVAL" in line 9 is the interval of file output. Positions and velocities of particles are written in output files. In this case, a file containing simulation results is written every 20 time steps. "ARRAY_SIZE" in line 10 expresses the size of an array and means the maximum total number of particles. In this case, 5,000 particles are available at the maximum. If you need to use more particles, you must increase the number of ARRAY_SIZE. "FINISH_TIME" in line 11 is the finish time of a simulation. In this case, after the simulation time reaches 2.0 s, the program ends.

2.3.4.2 Main Function

"main()" is the first function executed in a C language program. Program 2.6 is the main function of the sample program. The function in line 3 sets the initial coordinates and velocities of particles. For details of the initial setting, see the next section. The function in line 5 sets various constants used in MPS simulation. The function in line 6 is the main loop function of the simulation, in which procedures for a single time step are repeatedly executed.

```
001   int main (···) {
002      ...
003      initializeParticlePositionAndVelocity_for2dim();
004      ...
005      calculateConstantParameter();
006      mainLoopOfSimulation();
007      ...
008   }
```

Program 2.6

2.3.4.3 initializeParticlePositionAndVelocity_for 2dim( ) Function

This function sets up the initial coordinates and velocities of particles. Particles are generated at regular intervals in the simulation domain. The interval is l_0, which is the initial distance between particles and is expressed as PARTICLE_DISTANCE in the program. In the program, there are some strings placed between /* and */. The character strings between /* and */ are comment sentences and are not executed.

Note 2.24: *The "for" statement*: In line 4, there is a "for(...){...}" statement, which repeats the statements within the brace. For example, the following part of a program expresses a loop statement.

$$\text{for}(i = 1; i < = 3; i + +)"\{\text{printf("Counter is at \%d. n", }i)\}$$

where i = 1 is the initial value of i, "i < = 3"; means that the loop continues while i is less than or equal to 3. "i + +" means that i is increased by 1 after one loop completes. "printf(1stARGUMENT, 2ndARGUMENT)" expresses the 1stARGUMENT is written on your monitor. "%d" in the 1stARGUMENT expresses that the 2ndARGUMENT (in this case, the second argument is i) is written as a decimal integer. After executing the above program, the following lines are shown on your monitor.

Counter is at 1.

Counter is at 2.

Counter is at 3.

Lines 4—8 of Program 2.7 are statements for generating particles at even intervals from the position of the lower limit of the calculation domain. The variables x, y, and z express a candidate position, where a particle is going to be generated. In line 10, the candidate position is judged as to whether the position is in a fluid region or not. In this case, a fluid region exists in the range of $0\text{ m} < x \le 0.25\text{ m}$ and $0\text{ m} < y \le 0.5\text{ m}$. "EPS" is a very small value for correcting boundary positions and generating or not generating particles on a boundary.

In line 10, the particle type is set. In the sample program, the fluid, wall, dummy wall, and ghost particle type numbers are 0, 1, 2, -1, respectively. Dummy wall particles are located behind wall particles. The pressures of dummy wall particles are not calculated in the pressure Poisson equation because these pressures are assumed to be the same as the wall particles. The ghost particles do not affect the simulation at all because calculations involving ghost particles are skipped in the simulation program. The ghost particles are used to express an outflow. We can express an outflow by changing fluid particles to ghost particles. If the candidate position is in a wall region, the type number of the generating particle is set to 1. If the candidate position is not included in any regions, we do not generate a particle. The number "-4" in the program is for making four layers of boundary particles: two layers of wall particles and two layers of dummy wall particles. In line 19, the total number of particles is set. In line 20, the initial velocities of fluid particles are set. VelocityX[i], VelocityY[i], and VelocityZ[i] are the velocity components of the ith particle in x, y, and z directions, respectively. In this case, all fluid particles are initialized at 0.0 m/s. This means that the fluid is stationary at the beginning of the simulation.

```
001   void
002   initializeParticlePositionAndVelocity_for2dim( void ) {
003      ...
004      for (iX= -4; iX<nX; iX++) {
005         for (iY= -4; iY<nY; iY++) {
006            x = PARTICLE_DISTANCE * (double) (iX);
007            y = PARTICLE_DISTANCE * (double) (iY);
008            z = 0.0;
009            if( ((x>0.0)&&(x<=0.25)) && ((y>0.0)&&(y<=0.50)) ) {/*--- fluid region --*/
010               ParticleType[i]=0;
011               PositionX[i]=x; PositionY[i]=y; PositionZ[i]=z;
012               i++;
013            }else if(...) {/* -- wall region -- */
014               ...
015   }}}
016
017   NumberOfParticle = i;
018   for (i=0; i<NumberOfParticle*3; i++) {
019      VelocityX[i]=0.0; VelocityY[i]=0.0; VelocityZ[i]=0.0;
020   }
021   }
```

Program 2.7

2.3.4.4 calculateNZeroAndLambda( ) Function

This function calculates n^0 in Eq. (2.19), and λ^0 in Eq. (2.34). To calculate these parameters, particles are virtually arranged at regular intervals in x, y, and z directions in lines 11−17 of Program 2.8. In this program, each particle model can have their own effective radius of interaction zone. Therefore, we calculate each n^0 for the particle number density, gradient, and Laplacian models. In lines 8−9, we set n^0 to 0 for the initialization. In lines 15−20, the particle distance between the ith particle and jth particle is calculated. In lines 21−23, the weight of the jth particle is calculated by the weight function of Eq. (2.18). The calculated weight is added to n^0 to calculate the summation of weights. This calculation corresponds to Eq. (2.19). In lines 24 and 28, λ^0 is calculated on the basis of Eq. (2.34).

```
001   void calculateNZeroAndLambda( void ) {
002     ...
003     if( DIM == 2 ) {
004       iZ_start = 0; iZ_end = 1;
005     } else {
006       iZ_start = -4; iZ_end = 5;
007     }
008     N0_forNumberDensity = 0.0;  N0_forGradient   = 0.0;
009     N0_forLaplacian    = 0.0;  Lambda           = 0.0;
010     xi = 0.0;  yi = 0.0;  zi = 0.0;
011     for (iX= -4; iX<5; iX++) {
012       for (iY= -4; iY<5; iY++) {
013         for (iZ= iZ_start; iZ<iZ_end; iZ++) {
014           if( ((iX==0) && (iY==0)) && (iZ==0) ) continue;
015           xj = PARTICLE_DISTANCE * (double) (iX);
016           yj = PARTICLE_DISTANCE * (double) (iY);
017           zj = PARTICLE_DISTANCE * (double) (iZ);
018           distance2 = (xj-xi)*(xj-xi)+(yj-yi)*(yj-yi)+(zj-zi)*(zj-zi);
019           if( distance2 >= THRESHOLD) continue;
020           distance = sqrt(distance2);
021           N0_forNumberDensity += weight(distance, Radius_forNumberDensity);
022           N0_forGradient      += weight(distance, Radius_forGradient);
023           N0_forLaplacian     += weight(distance, Radius_forLaplacian);
024           Lambda  += distance2 * weight(distance, Radius_forLaplacian);
025         }
026       }
027     }
028     Lambda = Lambda/N0_forLaplacian;
029   }
```

Program 2.8

2.3.4.5 weight( ) Function

This function calculates the weight function of Eq. (2.18). The first argument, "distance," is the distance between particles. The second argument, "re," is the effective radius of the interaction zone. If the distance is longer or equal to the effective radius, we set the weight to zero. If not, we calculate the weight by Eq. (2.18). In line 8, the calculated weight is returned to the statement which called the function "weight()." (Program 2.9).

```
001   double weight(double distance, double re) {
002     double weightIJ;
003     if( distance >= re ) {
004       weightIJ = 0.0;
005     }else{
006       weightIJ = (re/distance) - 1.0;
007     }
008     return weightIJ;
009   }
```

Program 2.9

2.3.4.6 mainLoopOfSimulation( ) Function

This is the main loop function of a time step. In line 2, we set "iTimeStep," which is the counter of the time step, to 0 for the initialization. The statement, "while(1){ ... }" repeats the procedures inside the brace until the "break" command is executed. Lines 5—20 are a series of procedures for a time step. "calculateGravity()" in Line 6 sets the accelerations of particles to the acceleration of gravity. "calculateViscosity()" in Line 7 calculates the acceleration due to the viscosity term of the Navier—Stokes equation. "moveParticle()" in Line 8 updates velocities of particles and move particles based on the updated velocities. "calculatePressure()" in line 10 calculates the pressure of particles. "calculatePressureGradient()" in line 11 calculates the acceleration of particles due to the pressure gradient term of the Navier—Stokes equation. "moveParticlesUsing PressureGradient()" in line 12 corrects velocities and positions of particles. The above functions are major procedures for a time step. In line 14, the time is increased by Δt. The details of the above functions are explained later.

Lines 16—19 output the simulation results to a file. "%" in line 16 expresses the remainder of a division. " = = " expresses the left-hand side is equal to the right-hand side. In this case, an output file is written every 20 time steps because OUTPUT_INTERVAL is defined as 20.

Note 2.25: *Vtu file-format*: "writeData_inVtuFormat()" in line 17 outputs the simulation results in "vtu" format, which is a file format of ParaView and has the file extension of ".vtu." ParaView is free software for visualization. "writeData_inProfFormat()" in line 18 outputs the simulation results in "prof" format, whose file extension is ".prof." In a file with prof format, simulation time is written in the first line. The number of particles is written in the second line. From the third line onward, each particle's information is written. In each line, the program outputs x, y, and z coordinates, x, y, and z components of velocity, pressure, and the particle number density, respectively.

Note 2.26: *Prof file-format*: "In line 20, "Time," which is the simulation time, is checked against the finishing time. If "Time" reaches the finishing time, the "while( )" loop is terminated by the "break" command, and the simulation program ends. If "Time" has not reached the finishing time, the program returns to line 4, and continues the same procedures until the finishing time.

Note 2.27: *Collision*: The "collision()" function in line 8 is an exceptional procedure involving particle collisions. In particle simulations, particles rarely approach each other. As a result, the simulation becomes unstable. To avoid the instability, collisions between particles are simulated. In the case where pressure is implicitly calculated, particle collisions are very rare and are regarded as an exceptional procedure (Program 2.10).

```
001   void mainLoopOfSimulation( void ) {
002     int iTimeStep = 0;
003     ...
004     while(1) {
005       --
006       calculateGravity();
007       calculateViscosity();
008       moveParticle();
009       collision();
010       calculatePressure();
011       calculatePressureGradient();
012       moveParticleUsingPressureGradient();
013       iTimeStep++;
014       Time += DT;
015       ...
016       if( (iTimeStep%OUTPUT_INTERVAL) == 0 ) {
017         writeData_inVtuFormat();
018         writeData_inProfFormat();
019       }
020       if(Time >= FINISH_TIME ) {break;}
021     }
022   }
```

Program 2.10

2.3.4.7 calculateGravity Function

This function sets the array of particle acceleration to the acceleration of gravity (the third term of the Navier–Stokes equations Eq. 2.6). The variable "i" expresses a particle ID. "NumberOfParticle" expresses the total number of particles. The "for" loop in line 4 repeats the procedures changing the value of i for all particles. The elements AccelerationX[i], AccelerationY[i], and AccelerationZ[i] express the x-, y-, and z-directional components of the ith particle's acceleration. If the ith particle is a fluid particle, we set these elements to the acceleration of gravity. If the ith particle is not a fluid particle, we set these elements to $0 \, m/s^2$ because the ith particle is a wall particle, dummy wall particle, or ghost particle, which is stationary (Program 2.11).

```
001   void calculateGravity( void ) {
002     int i;
003
004     for(i=0;i<NumberOfParticle;i++) {
005       if(ParticleType[i] == FLUID) {
006         AccelerationX[i]=GRAVITY_X;
007         AccelerationY[i]=GRAVITY_Y;
008         AccelerationZ[i]=GRAVITY_Z;
009       }else{
010         AccelerationX[i]=0.0;
011         AccelerationY[i]=0.0;
012         AccelerationZ[i]=0.0;
013       }
014     }
015   }
```

Program 2.11

2.3.4.8 calculateViscosity Function

This function calculates the acceleration due to the viscosity, which is the second term of the right-hand side of Eq. (2.46). The parameter i is the ID of a particle. Variables viscosityTermX, viscosityTermY, and viscosityTermZ express the x, y, and z components of acceleration due to the viscous force, respectively. Line 6 in Program 2.12 sets components to $0 \, m/s^2$ for the initialization.

Note 2.28: *How to calculate $\sum_{j \neq i}$*: The "for" statement in line 7 repeats the same procedures changing the jth particle. The additions in lines 16−18 and the repeating statement "for" in line 7 express $\sum_{j \neq i}$, which is the summation calculation of Eq. (2.46). In line 8, the "continue" command is executed. The "continue" command skips the statements that follow in the "for" loop. The "continue" command in line 8 expresses $j \neq i$ of $\sum_{j \neq i}$, which means the ith particle is excluded from the summation of the jth particle.

The, "continue" statement is executed in the case where ParticleType [j] is equal to GHOST. This avoids calculation of the viscosity between a fluid particle and a ghost particle j. In lines 9−14, the distance between the ith and jth particles is calculated. "sqrt" calculates the square root of the argument. In lines 14−19, the weight of the jth particle is calculated and is used for the calculation of $\sum_{j\neq i}\left(\mathbf{u}_j - \mathbf{u}_i\right)w\left(\left|\mathbf{r}_j - \mathbf{r}_i\right|, r_e\right)$ in Eq. (2.46). The velocity vector $\mathbf{u}$ has x-, y-, and z-directional components. Therefore, in lines 16−18, the viscosity calculation is carried out for each directional component of velocity. The right-hand sides of lines 21−23 are calculations for multiplying $\frac{2d}{\lambda^0 n^0}$ by the result of the above summation. In lines 24−26, we add the calculated acceleration due to viscosity to the array of the acceleration of a fluid particle i. By the above calculation, we can obtain the acceleration due to viscosity of the ith particle. We repeat the above procedure by changing i in the "for" statement in line 4.

```
001    void calViscosity( void ) {
002      ...
003      a = (2.0*KINEMATIC_VISCOSITY*DIM)/(NO_forLaplacian*Lambda);
004      for(i=0;i<NumberOfParticle;i++){
005        if(ParticleType[i] != FLUID) continue;
006        viscosityTermX=0.0; viscosityTermY=0.0; viscosityTermZ=0.0;
007        for(j=0;j<NumberOfParticle;j++){
008          if( (j==i) || (ParticleType[j]==GHOST) ) continue;
009          xij = PositionX[j] - PositionX[i];
010          yij = PositionY[j] - PositionY[i];
011          zij = PositionZ[j] - PositionZ[i];
012          distance2 = (xij*xij) + (yij*yij) + (zij*zij);
013          if(distance2 >= THRESHOULD) continue;
014          distance = sqrt(distance2);
015          w =  weight(distance, Radius_forLaplacian);
016          viscosityTermX +=(VelocityX[j]-VelocityX[i])*w;
017          viscosityTermY +=(VelocityY[j]-VelocityY[i])*w;
018          viscosityTermZ +=(VelocityZ[j]-VelocityZ[i])*w;
019        }
020
021        viscosityTermX = viscosityTermX * a;
022        viscosityTermY = viscosityTermY * a;
023        viscosityTermZ = viscosityTermZ * a;
024        AccelerationX[i] += viscosityTermX;
025        AccelerationY[i] += viscosityTermY;
026        AccelerationZ[i] += viscosityTermZ;
027      }
028    }
029
030
```

Program 2.12

2.3.4.9 moveParticle( ) Function

This function updates the particle positions. In lines 6−8, velocities of fluid particles are updated by the particle accelerations, which consist of the accelerations due to gravity and viscosity. "VelocityX[i], VelocityY[i], and VelocityZ[i]" are velocity components in x, y, and z directions, respectively. In the same manner, "PositionX[i], PositionY[i], and PositionZ[i]" are x, y, and z coordinates of the ith particle. "DT" is the time increment Δt and is set to 0.001 s. In lines 13−15, the acceleration components are reset to 0.0 m/s^2 (Program 2.13).

```
001   void moveParticle( void ) {
002     int i;
003
004     for(i=0;i<NumberOfParticle;i++){
005       if(ParticleType[i] == FLUID){
006         VelocityX[i] +=AccelerationX[i]*DT;
007         VelocityY[i] +=AccelerationY[i]*DT;
008         VelocityZ[i] +=AccelerationZ[i]*DT;
009         PositionX[i] +=VelocityX[i]*DT;
010         PositionY[i] +=VelocityY[i]*DT;
011         PositionZ[i] +=VelocityZ[i]*DT;
012         AccelerationX[i]=0.0;
013         AccelerationY[i]=0.0;
014         AccelerationZ[i]=0.0;
015       }
016     }
017   }
```

Program 2.13

2.3.4.10 calculatePressure( ) Function

This function calculates pressure. The implicit algorithm explained in Section 2.2.5.1 is used in this function. In line 2, the function "calculateParticleNumberDensity()" calculates the particle number density, which is used in the right-hand side of Eq. (2.49). In line 3, "setBoundaryCondition()" sets the Dirichlet boundary condition of the pressure Poisson equation. In line 4, "setSourceTerm()" sets the right-hand side of the pressure Poisson equation. In line 5, "setMatrix()" creates the coefficient matrix **A** of the pressure Poisson equation.

The above procedures express the pressure Poisson equation (Eq. (2.49)) as the simultaneous linear equation of Eq. (2.77). The simultaneous liner equations are solved by Gaussian elimination in line 6. Although we adopted Gaussian elimination because of its simplicity, the speed of the solver is slow. The calculation amount of Gaussian elimination is in proportion to N^3, where N is the number of unknown parameters. In this case, the unknown parameters are the pressures of particles. Therefore, the calculation amount becomes very large if the number of particles is large. The coefficient matrix is symmetrical and sparse. For these the characteristics, we should use a faster solver, such as the conjugate gradient method, for solving the simultaneous equations.

In line 7, "removeNagativePressure()" resets negative pressures to 0 Pa, which is the same as the pressure of a free surface. Around a free surface, the particle number of density n^* tends to be lower than n^0 because there are no particles over a free surface. As a result, negative pressures often occur around the free surface. To avoid fluctuation and instability due to the negative pressures, if the pressure of a particle is lower than that of the free surface, we reset the particle's pressure to that of the free surface. The function "setMinimumPressure()" calculates the minimum pressure around each particle. The minimum pressure is used in the calculation of the pressure gradient explained later (Program 2.14).

```
001    void calculatePressure( void ) {
002      calculateNumberDensity();
003      setBoundaryCondition();
004      setSourceTerm();
005      setMatrix();
006      solveSimultaniousEquationsByGaussEliminationMethod();
007      removeNegativePressure();
008      setMinimumPressure();
009    }
```

Program 2.14

2.3.4.11 calculateNumberDensity( ) Function

This function calculates the particle number densities defined in Eq. (2.17). In lines 8−13, the distance between particle i and a neighboring particle j is calculated. In line 14, the weight of particle j is calculated using the calculated distance between particles. In line 15, the obtained weight is added to "ParticleNumberDensity[i]," which is the array element for saving the particle number density of the ith particle (Program 2.15).

```
001   void calculateParticleNumberDensity( void ) {
002     ...
003     for(i=0; i<NumberOfParticle; i++) {
004       if(ParticleType[i] != GHOST) {
005         NumberDensity[i] = 0.0;
006         for(j=0; j<NumberOfParticle; j++) {
007           if( (j==i) || (ParticleType[j]==GHOST) ) continue;
008           xij = PositionX[j] - PositionX[i];
009           yij = PositionY[j] - PositionY[i];
010           zij = PositionZ[j] - PositionZ[i];
011           distance2 = (xij*xij) + (yij*yij) + (zij*zij);
012           if( distance2 >= THRESHOLD ) continue;
013           distance = sqrt(distance2);
014           w = weight(distance, Radius_forNumberDensity);
015           ParticleNumberDensity[i] += w;
016         }
017       }
018     }
    }
```

Program 2.15

2.3.4.12 setBoundaryCondition( ) Function

This function sets the Dirichlet boundary condition of the pressure Poisson equation. Specifically, we set flags, which express whether each pressure is fixed or not. By the "for" statement in line 3, the procedures in the "for" loop are repeatedly carried out. In lines 4—5, if the ith particle is a dummy or ghost particle, we set the flag to "GHOST_OR_DUMMY" which skips the pressure calculation of the particle because the pressure of a ghost or dummy wall particle is not used in the simulation. In lines 6—7, if the particle number density of the ith particle is lower than a certain value, the particle is regarded as a free-surface particle, and the flag for the boundary condition is set to "SURFACE_PARTICLE." For other cases, we set the flag to "INNER_PARTICLE" because we need to determine the particle pressure. These flags are used to solve the simultaneous linear equations with respect to pressure (Program 2.16).

```
001   void setBoundaryCondition( void ) {
002     ...
003     for(i=0; i<NumberOfParticle; i++) {
004       if(ParticleType[i]==GHOST || ParticleType[i]== DUMMY_WALL ) {
005         BoundaryCondition[i]=GHOST_OR_DUMMY;
006       }else if( ParticleNumberDensity[i]/n0 < THRESHOLD_RATIO_OF_NUMBER_DENSITY) {
007         BoundaryCondition[i]=SURFACE_PARTICLE;
008       }else{
009         BoundaryCondition[i]=INNER_PARTICLE;
010       }
011     }
012   }
```

Program 2.16

2.3.4.13 setSourceTerm( ) Function

This function sets the right-hand side of the simultaneous linear equations with respect to pressure. Line 6 is the calculation of Eq. (2.59) (Program 2.17).

```
001   void setSourceTerm( void ) {
002     ...
003     for(i=0;i<NumberOfParticle;i++) {
004       if(ParticleType[i]==GHOST || ParticleType[i]== DUMMY_WALL ) continue;
005       if(BoundaryCondition[i]==INNER_PARTICLE) {
006         SourceTerm[i] = (1.0/(DT*DT))*((ParticleNumberDensity[i]-n0)/n0);
007       }else if(BoundaryCondition[i]==SURFACE_PARTICLE) {
008         SourceTerm[i]=0.0;
009       }
010     }
011   }
```

Program 2.17

2.3.4.14 setMatrix( ) Function

This function sets up the coefficient matrix **A** of the left-hand side of the simultaneous linear equations with respect to pressure (Eq. 2.77). In lines 8−13, we calculate the distance between particle i and a neighboring particle j. Using the calculated distance, each element of the coefficient matrix, which is expressed in Eq. (2.8), is calculated, where CoefficientMatrix[i][j] is an element a_{ij} of Eq. (2.73). Line 19 is the calculation of the compressibility. For details, see Section 3.1.

Note 2.29: *Exceptional procedure for pressure calculation*: The function in line 21 is for an exceptional procedure. A fluid particle group, which does not have any Dirichlet boundary conditions, rarely occurs. For example, if two isolated fluid particles approach each other and the weight function can become large enough that both fluid particles can become inner particles. Those fluid groups do not have any free-surface particles because every particle has a larger particle number density than the threshold of free-surface particles. If such an exceptional situation occurs, we increase the diagonal elements of the coefficient matrix **A** of the pressure Poisson equation. In this case, we double the diagonal elements of the particle groups which do not have any Dirichlet boundary conditions. By increasing these elements, we can obtain the solution of the simultaneous linear equations even if there are particle group which do not have Dirichlet boundary conditions because increasing diagonal terms create a situation in which virtual fluid particles of 0 Pa pressure exist around the particle groups (Program 2.18).

```
001   void setMatrix( void ) {
002       ...
003       a = 2.0*DIM/(n0*Lambda) ;
004       for(i=0; i<NumberOfParticle; i++) {
005           if(BoundaryCondition[i] != INNER_PARTICLE) continue;
006           for(j=0; j<NumberOfParticle; j++) {
007               if( (j==i) || (ParticleType[j]==GHOST) ) continue;
008               xij = PositionX[j] - PositionX[i];
009               yij = PositionY[j] - PositionY[i];
010               zij = PositionZ[j] - PositionZ[i];
011               distance2 = (xij*xij)+(yij*yij)+(zij*zij) ;
012               if(distance2 >= THRESHOLD) continue;
013               distance  = sqrt(distance2) ;
014               if(distance>=Radius_forLaplacian)continue;
015               coefficientIJ = a * weight(distance, Radius_forLaplacian)/FluidDensity;
016               CoefficientMatrix[i*NumberOfParticle+j]  = coefficientIJ;
017               CoefficientMatrix[i*NumberOfParticle+i] -= coefficientIJ;
018           }
019           CoefficientMatrix[i*NumberOfParticle+i] -= COMPRESSIBILITY/(DT*DT) ;
020       }
021       exceptionalProcessingForBoundaryCondition() ;
022   }
```

Program 2.18

2.3.4.15 solveSimultaniousEquationsByGaussianElimination( ) Function

This function solves the simultaneous linear equations (Eq. 2.77) on the basis of the Gaussian elimination and obtains the particle pressures. In line 3, the initial solution is set to 0 Pa. The initial solution is also used as the constant pressure of free-surface particles. Lines 6–16 are procedures for the forward elimination of the Gaussian elimination, while lines 18–26 are procedures for the backward substitution. "CoefficientMatrix[i][j]" expresses a_{ij} of Eq. (2.73). In lines 7 and 19, we skip calculations for free-surface particles, ghost particles, and dummy-wall particles by using the "continue" statement because pressures of these particles are either known or not used (Program 2.19).

```
001   void solveSimultaniousEquationsByGaussianElimination ( void ) {
002     --
003     int n=NumberOfParticle;
004     for(i=0; i<n; i++) { Pressure[i] = 0.0; }
005
006     for(i=0; i<n-1; i++) {
007       if ( BoundaryCondition[i] != INNER_PARTICLE ) continue;
008       for(j=i+1; j < n; j++) {
009         alpha = CoefficientMatrix[j][i]/CoefficientMatrix[i][i];
010         for(j=i+1; j < NumberOfParticle; j++) {
011         CoefficientMatrix[j][j]=CoefficientMatrix[j][j]
012                             alpha * CoefficientMatrix[i][j];
013         }
014         SourceTerm[j] = SourceTerm[j] - alpha*SourceTerm[i];
015       }
016     }
017
018     for( i=n-1; i>=0; i--) {
019       if ( BoundaryCondition[i] != INNER_PARTICLE ) continue;
020       sumOfTerms = 0.0;
021       for( j=i+1; j<n; j++ ) {
022         sumOfTerms += CoefficientMatrix[i][j] * Pressure[j];
023       }
024       Pressure[i] = (SourceTerm[i] - sumOfTerms)
025                     /CoefficientMatrix[i][i];
026     }
027   }
```

Program 2.19

2.3.4.16 calculatePressureGradient( ) Function

This function (Program 2.20) calculates the acceleration vectors due to the pressure gradient term, which is the second term on the right-hand side of Eq. (2.50). The gradient model of the MPS method is applied to the first derivative of pressure as follows:

$$-\frac{1}{\rho^0}\langle\nabla P\rangle_i^{k+1} = -\frac{1}{\rho^0}\frac{d}{n^0}\sum\left[\frac{P_j^{k+1}-P_i^{k+1}}{\left|\mathbf{r}_j^*-\mathbf{r}_i^*\right|^2}\left(\mathbf{r}_j^*-\mathbf{r}_i^*\right)w\left(\left|\mathbf{r}_j^*-\mathbf{r}_i^*\right|,r_e\right)\right]$$

(2.82)

The above gradient model is a little different from Eq. (2.24). We use $\hat{P}_i$ instead of the pressure of the ith particle, P_i. The variable $\hat{P}_i$ is the minimum pressure among ith particle and its neighborhood particles. By using the minimum pressure, $P_j - \hat{P}_i$ in Eq. (2.82) is positive or zero. As a result, attractive force do not work between the ith and jth particles, and

```
001   void calPressureGradient( void ){
002       ...
003       double a;
004       a =DIM/NO_forGradient;
005
006       for(i=0;i<NumberOfParticle;i++) {
007           if(ParticleType[i] != FLUID) continue;
008           gradientX = 0.0; gradientY = 0.0; gradientZ = 0.0;
009           for(j=0;j<NumberOfParticle;j++) {
010               if( (j==i) || (ParticleType[j]==GHOST) ) continue;
011               xij = PositionX[j] - PositionX[i];
012               yij = PositionY[j] - PositionY[i];
013               zij = PositionZ[j] - PositionZ[i];
014               distance2 = (xij*xij) + (yij*yij) + (zij*zij);
015               if(distance2 >= THRESHOLD) continue;
016               distance = sqrt(distance2);
017               if(distance<Radius_forGradient) {
018               w = weight(distance, Radius_forGradient);
019               pij = (Pressure[j] - MinimumPressure[i])/distance2;
020               gradientX += xij*pij*w;
021               gradientY += yij*pij*w;
022               gradientZ += zij*pij*w;
023               }
024           }
025           gradientX *= a;
026           gradientY *= a;
027           gradientZ *= a;
028           AccelerationX[i]= (-1.0)*gradientX/FluidDensity;
029           AccelerationY[i]= (-1.0)*gradientY/FluidDensity;
030           AccelerationZ[i]= (-1.0)*gradientZ/FluidDensity;  }
031   }
```

Program 2.20

we can restrain the instability due to attractive forces. In the derivation of Eq. (2.82), we introduced the following approximation, which is satisfied in the case where particles are distributed uniformly and isotopically.

$$\frac{1}{n^0} \sum \left[\frac{(\mathbf{r}_j - \mathbf{r}_i)}{|\mathbf{r}_j - \mathbf{r}_i|^2} w\left(|\mathbf{r}_j - \mathbf{r}_i|, r_e \right) \right] \cong \mathbf{0} \qquad (2.83)$$

The right-hand side is the zero vector. In lines $11-16$, the distance between ith and jth particles is calculated. In lines $18-22$, the summation of Eq. (2.82) is calculated.

2.3.4.17 calculatePressure_forExplicitMPS( ) Function

This is a function for the explicit MPS method. This function (Program 2.21) calculates pressure on the basis of the equation of state (Eq. 3.38 in Section 3.2) instead of the pressure Poisson equation. The equation of

```
001  void calculatePressure_forExplicitMPS( void ) {
002
003     int     iParticle;
004     double n0;
005     double ni;
006     double c, c2;
007     double rho = FLUID_DENSITY;
008
009     c  = SPEED_OF_SOUND;
010     c2 = c * c;
011     n0 = NO_forParticleNumberDensity;
012
013     for(iParticle=0; iParticle < NumberOfParticles; iParticle++) {
014        if ( (ParticleType[iParticle]==GHOST)
015           || (ParticleType[iParticle]==DUMMY_WALL) ) {
016          Pressure[iParticle] = 0.0;
017        }else{
018          ni  = ParticleNumberDensity[iParticle];
019          Pressure[iParticle] =  c2 * rho * (ni - n0) / n0;
020          If( Pressure[iParticle] < 0.0 ){
021             Pressure[iParticle] =  0.0;
022          }
023        }
024     }
025
026  }
027
028
029
```

Program 2.21

state is expressed by the statement in line 20. To avoid instability due to negative pressures, negative pressures are set to 0 Pa in lines 20−22 of this program.

2.3.4.18 calculatePressureGradient_forExplicitMPS( ) Function

This is a function for the explicit MPS method. The function (Program 2.22) calculates the acceleration vectors of the pressure gradient term. The processing is similar to Program 2.20, which is a function for the MPS method. There is a difference between the two programs in line 19. Eqs. (3.43) and (3.44) are used to discretize the pressure gradient in the explicit MPS method, while Eq. (2.82) is used in the MPS method. The gradient model of Eqs. (3.43) and (3.44) can increase the repulsive force between particles and stabilize the simulation.

```
001   void calculatePressureGradient_forExplicitMPS( void ) {
002     ...
003     double a;
004     a =DIM/NO_forGradient;
005
006     for(i=0;i<NumberOfParticle;i++) {
007       if(ParticleType[i] != FLUID) continue;
008       gradientX = 0.0; gradientY = 0.0; gradientZ = 0.0;
009       for(j=0;j<NumberOfParticle;j++) {
010         if( (j==i) || (ParticleType[j]==GHOST) ) continue;
011         xij = PositionX[j] - PositionX[i];
012         yij = PositionY[j] - PositionY[i];
013         zij = PositionZ[j] - PositionZ[i];
014         distance2 = (xij*xij) + (yij*yij) + (zij*zij);
015         if(distance2 >= THRESHOLD) continue;
016         distance = sqrt(distance2);
017         if(distance<Radius_forGradient) {
018         w = weight(distance, Radius_forGradient);
019         pij = (Pressure[j] + Pressure[i])/distance2;
020         gradientX += xij*pij*w;
021         gradientY += yij*pij*w;
022         gradientZ += zij*pij*w;
023         }
024       }
025       gradientX *= a;
026       gradientY *= a;
027       gradientZ *= a;
028       AccelerationX[i]= (-1.0)*gradientX/FluidDensity;
029       AccelerationY[i]= (-1.0)*gradientY/FluidDensity;
030       AccelerationZ[i]= (-1.0)*gradientZ/FluidDensity;  }
031   }
```

Program 2.22

2.4 EXERCISE OF SIMULATION

Let experience MPS simulations though exercises. We can solve the following exercises using the sample programs of this book.

2.4.1 Exercises

Exercise 1: Download the sample program of the MPS method (mps.c). Then, compile and execute the program.

Exercise 2: Visualize the simulation results of Exercise 1 with ParaView. Then, color the pressure distribution. After that, color the velocity distribution.

Exercise 3: Edit "mps.c" and reduce the spatial resolution, which is expressed by the initial distance between particles, to 0.05 m.

Then, compile and execute the revised program. Finally, visualize the simulation results with ParaView.

Exercise 4: Edit "mps.c" and increase the kinematic viscosity to 0.1 m^2/s. Then, compile and execute the revised program. Finally, visualize the simulation results with ParaView.

Exercise 5: Edit "mps.c" and change the initial position of the fluid region to the central region (0.4 m $<x<$ = 0.6 m, 0 m $<y<$ = 0.5 m). Then, compile and execute the revised program. Visualize the simulation results with ParaView.

Exercise 6: Download the sample program of the explicit MPS method (emps.c). Then, compile and execute the program. Finally, visualize the simulation results with ParaView.

Exercise 7: Edit "emps.c" and change the number of spatial dimensions to 3 to carry out a three-dimensional simulation. Then, compile and execute the revised program. Finally, visualize the simulation results with ParaView.

2.5 HINTS FOR EXERCISES

Hint for exercise 1: See the beginning of Sections 2.3 and 2.3.2

Hint for exercise 2: See Section 2.3.3

Hint for exercise 3: Revise line 14 in the sample program (mps.c) as follows:

[Before revision]

```
#define PARTICLE_DISTANCE 0.025
```

[After revision]

```
#define PARTICLE_DISTANCE 0.050
```

Hint for exercise 4: Revise line 28 in the sample program (mps.c) as follows:

[Before revision]

```
#define KINEMATIC_VISCOSITY (1.0E-6)
```

[After revision]

```
#define KINEMATIC_VISCOSITY (1.0E-1)
```

Hint for exercise 5: Revise line 161 in the sample program (mps.c) as follows:

[Before revision]

```
if(    ((x > 0.0 + EPS)&&(x < = 0.25 + EPS))&&((y > 0.0 + EPS)&&
(y < = 0.50 + EPS)) ){
```

[After revision]

if(((x > 0.0 + EPS)&&(x < = 0.40 + EPS))&&((y > 0.0 + EPS)&&
(y < = 0.50 + EPS))){

Answer for exercise 6: Revise line 13 in the sample program (emps.c) as follows:

    [Before revision]
    #define DIM 2
    [After revision]
    #define DIM 3

2.6 FREQUENTLY ASKED QUESTIONS

This section describes frequently asked questions and the answers.

2.6.1 What Is the Best Effective Radius of the Interaction Zone?

In the above examples, we used a relatively short effective radius of $1.5l_0$. That was for simplifying the explanation of the models because we can reduce the number of neighboring particles if we adopt a short effective radius. Typically, we use $2.0l_0$ to $4.1l_0$ for the effective radius because the approximation of uniform and isotropic particle distribution is improved by using a long effective radius, and simulations are stabilized. Although a long radius might stabilize simulations, we should not use too long a radius because it increases the number of neighboring particles, which lengthens the computation time. Moreover, too long a radius reduces the accuracy of approximations because it is difficult to approximate the physical quantity of a local region. In particular, the number of neighboring particles in three-dimensional simulations is more than that in two-dimensional simulations even if effective radii are the same in both dimensions. As a result, the computation time of three-dimensional simulations is longer than two-dimensional simulations. Although a too small or too large effective radius has a bad influence on simulations, the simulation results do not significantly depend on the effective radius if the radius is within a certain range, such as the range of $2.0l_0$ to $4.1l_0$.

2.6.2 Why Do We Need to Arrange Dummy Wall Particles Behind Wall Particles?

That is to appropriately assign a boundary condition to a wall boundary. By using wall particles, we can prevent penetration of fluid particles through a wall. Wall particles are located at a place where fluid directly contacts with. Each wall particle has a pressure variable and repels fluid particles by the pressure gradient term. On the other hand, dummy wall particles are located behind wall particles and have no contact with fluid particles. Dummy wall particles are used only for calculating the particle number densities of wall particles and fluid particles. If there are no dummy wall particles, most wall particles will have lower particle number densities than the threshold for judging free-surface particles (see Eq. 2.81). With dummy wall particles, we can avoid assigning an incorrect Dirichlet boundary condition to wall boundaries.

The pressures of dummy wall particles are not used for the pressure gradient calculations. Specifically, in the calculation of the pressure Poisson equation and the pressure gradient calculation, if a neighboring particle j is a dummy wall particle, the particle interaction between the ith and jth particle is not calculated. The exclusion of a dummy wall particle corresponds to the condition where the pressure of dummy wall particle j is equal to that of the ith particle. Therefore, by using dummy wall particles, we can approximately express the Neumann boundary condition where the pressure gradient is zero.

Pressures of wall particles are calculated in the same manner as fluid particles. We usually set the density of wall particles to that of fluid particles. The only difference between wall particles and fluid particles is that wall particles are fixed, and do not move, while fluid particles are moved by forces acting on them.

2.6.3 Particles Penetrated a Wall: What Is the Possible Reason?

One of the possible reasons is that wall particles and dummy wall particles are not arranged adequately. In typical MPS simulations, we usually use the effective radius of $2.1l_0$ for the interaction zone of the particle number density and the pressure gradient. In these cases, it is desirable to use double layers of wall particles and double layers of dummy wall particles to express a wall boundary.

2.6.4 How Do We Set the Time Increment Δt?

We need to use a sufficiently small time increment Δt that satisfies the CFL (Courant–Friedrich–Lewy) condition explained in Section 3.5 (Eqs. 3.51 and 3.52) and the diffusion number. However, small Δt

increases the necessary time steps and the computation time. Therefore, we should not use too small a Δt.

2.6.5 It Seems a Simulation Diverged Because Particles Exploded: What Is the Reason?

One of the possible reasons is that variables required for calculation are not correctly set up. We need to check whether Δt satisfies the the CFL condition and the diffusion number. We also need to check whether the initial distance between particles l^0 is correct. For example, we need to confirm that particles are arranged at even intervals in the initial state.

2.6.6 Fluid Was Compressed: What Is the Reason?

Variables such as l^0, n^0, and λ^0 might not be set correctly. Make sure that these parameters are correct. One of the other possible reasons is that two particles mistakenly exist at the same position in the initial configuration. If the distance between particles mistakenly becomes zero, the weight function of the MPS method (Eq. 2.18) is calculated as infinity due to the zero in the denominator, and problems occur.

2.6.7 How Can We Add a Function for Inlet or Outlet Boundaries in a Simulation Program?

Outlet boundaries are expressed by deleting fluid particles. We judge whether a particle flows out or not by its coordinate. If a fluid particle exits the simulation domain, we change the fluid particle to a ghost particle, which does not affect simulations. We can add a function for an outlet boundary in the sample program by adding the following command:

$$\text{ParticleType[i]} = -1;$$

where "-1" indicates that the particle type is a ghost, and i expresses the ID of the fluid particle.

On the other hand, we can express an inlet by changing a ghost particle i to a fluid particle as follows:

$$\text{ParticleType[i]} = 0;$$

where "0" indicates that the particle type is "fluid." Of course, we need to determine the position and velocity of the generating particle. For the details of inlets and outlets, see Section 4.4.

2.6.8 What Is the Most Time-Consuming Part in an MPS Simulation?

In the case where the semi-implicit method is applied and the number of particle is relatively large, the pressure calculation is the most time consuming because it needs to solve simultaneous linear equations to calculate pressure. The dimensions of the coefficient matrix $\mathbf{A}$ of the simultaneous linear equations are $N \times N$, where N is the number of particles. If we apply the conjugate gradient method for solving the simultaneous equations, the calculation amount is in proportion to about $N^{1.5}$ in two-dimensional simulations. If we apply an explicit method, the calculation amount for pressure calculation is in proportion to N.

The second most time-consuming part is searching for neighboring particles, although the function is not included in the sample program. We usually make a list of neighboring particles at the beginning or end of a time step. The list allows us to efficiently calculate the summations which appear in particle models (e.g., Eqs. 2.17, 2.24, and 2.33), because neighboring particles j are restricted by the list. It takes a long time to construct the list of neighboring particles. To reduce the computation time, a bucket method, which is a domain decomposition technique, is usually applied.

2.6.9 What Are the Drawbacks and the Strong Points of the Semi-implicit Method?

The semi-implicit method needs to solve simultaneous linear equations with respect to pressure. Therefore, the computation time per time step of the semi-implicit method is longer than an explicit method. However, the total computational time of the semi-implicit method is not always longer than the explicit method because the semi-implicit method can use a relatively large time increment Δt even if the fluid flow is incompressible. A large Δt allows us to reduce the required number of time steps. Moreover, the semi-implicit method does not need to adjust complicated variables, such as the speed of sound, which is often adjusted in the explicit method to relax the restriction regarding Δt.

REFERENCES

Hughes, W., Brighton, J., 1999. Schaum's Outline of Fluid Dynamics (Schaum's). McGraw-Hill Education.

Ikeda, H., 1999. Numerical Analysis of Fragmentation Processes in Vapor Explosions Using Particle Method. Dissertation of the University of Tokyo.

Kaye & Laby. <http://www.kayelaby.npl.co.uk/>.

Koshizuka, S., Oka, Y., 1996. Moving-particle semi-implicit method for fragmentation of incompressible fluid. Nucl. Sci. Eng. 123 (3), 421–434.

Potter, M., Wiggert, D.C., 2007. Schaum's Outline of Fluid Mechanics (Schaum's Outline Series). McGraw-Hill Education.

Tamai, T., Koshizuka, S., 2014. Least squares moving particle semi-implicit method. Comput. Particle Mech. 1, 277–305.

Tanaka, M., Masunaga, T., 2010. Stabilization and smoothing of pressure in MPS method by quasi-compressibility. J. Comput. Phys. 229 (11), 4279–4290.

CHAPTER 3

Extended Algorithms

Abstract

Two compressible-incompressible unified algorithms considering slight and strong compressibility are explained. An explicit algorithm using pseudo-compressibility is shown for the Moving Particle Semi-implicit (MPS) method. Numerical stability is compared for the explicit and semi-implicit algorithms. Symplectic schemes based on Hamiltonian system are developed for mechanical energy conservation. RATTLE can be used for incompressible flow. Arbitrary Lagrangian-Eulerian (ALE) approach is provided with the techniques of meshless discretization of the convection terms. A rigid body model using a quaternion and an algorithm to analyze interaction between the fluid and the rigid bodies are presented. A brief explanation of the elastic solid analysis using particles is shown. Calculation examples are also provided for these extended algorithms.

Keywords: Compressible-incompressible unified algorithm; pseudo-compressibility; symplectic scheme; mechanical energy conservation; Arbitrary Lagrangian-Eulerian (ALE); rigid body; quaternion; fluid—rigid body interaction; structural analysis; fluid—structure interaction

Contents

3.1 Compressible-Incompressible Unified Algorithm 111
3.2 Explicit Algorithm Using Pseudo-Compressibility 118
3.3 Symplectic Scheme 125
3.4 Arbitrary Lagrangian-Eulerian 134
3.5 Rigid Body Model 140
3.6 Structural Analysis 146
References 150
Further Reading 153

3.1 COMPRESSIBLE-INCOMPRESSIBLE UNIFIED ALGORITHM

The initial Moving Particle Semi–implicit (MPS) method (Koshizuka and Oka, 1996) employed the pressure Poisson equation for incompressibility:

$$\left\langle \nabla^2 P^{k+1} \right\rangle_i = -\frac{\rho}{\Delta t^2} \frac{n_i^* - n^0}{n^0}. \tag{3.1}$$

A little modification enables us to analyze slight linear compressibility as shown in Koshizuka et al. (1999) and Ikeda et al. (2001):

$$\left\langle \nabla^2 P^{k+1} \right\rangle_i = -\frac{\rho}{\Delta t^2} \frac{n_i^* - n_i^{k+1}}{n^0}, \tag{3.2}$$

where,

$$n_i^{k+1} = n^0 + \frac{n^0}{\rho} \frac{\partial \rho}{\partial P} P_i^{k+1}. \tag{3.3}$$

The particle number density at new-time step $k+1$, n_i^{k+1}, is not the initial value n^0 but a linear function of pressure P_i^{k+1} as shown in Eq. (3.3). Eq. (3.2) can be rewritten as

$$\left\langle \nabla^2 P^{k+1} \right\rangle_i = -\frac{\rho}{\Delta t^2} \left\{ \frac{n_i^* - n^0}{n^0} - \alpha P_i^{k+1} \right\}. \tag{3.4}$$

where α stands for compressibility

$$\alpha = \frac{1}{\rho c^2} = \frac{1}{\rho} \frac{\partial \rho}{\partial P}, \tag{3.5}$$

and c denotes the sound speed. This formulation is called slightly compressible MPS.

When α is zero, Eq. (3.4) is just the same as Eq. (3.1). This means that the fluid is perfectly incompressible. When α has a small value the particle number density is proportional to pressure. This means that the fluid is compressible and the relation between the density and the pressure is linear. However, the constant fluid density ρ is assumed in the Navier-Stokes equations. So the density change should be slight in this formulation.

The left hand side of Eq. (3.4) is discretized to a matrix using the Laplacian model in the MPS method. In this matrix the diagonal element is the same as the summation of the nondiagonal elements. The first term of the right hand side is the same as the source term of that of the perfectly incompressible fluid. The second term $(\rho/\Delta t^2) \alpha P_i^{k+1}$ is moved to the left hand side and added to the diagonal element in the matrix. This leads to a diagonal dominant matrix. In general, convergence of the iterative solvers is faster for the diagonal dominant matrix. Physically the source term of the pressure Poisson equation represents the temporary deviation from the constant density of the fluid. To solve the pressure Poisson equation is to resolve the deviation from the constant fluid density and to recover to the constant fluid density. In a perfectly incompressible fluid the deviation of the fluid density cannot be resolved inside the fluid and can be resolved on the boundary. Mathematically, this situation is represented by the matrix of which the diagonal element has the same value as the summation of the nondiagonal elements. On the other hand, in a slightly compressible fluid with a small value of compressibility, the deviation of the fluid density can be resolved inside the fluid as well as on the boundary. Mathematically, this situation is represented by the diagonal dominant matrix. So, convergence is expected to be faster for the slightly compressible fluid than for the perfectly incompressible fluid.

When Eq. (3.4) is used instead of Eq. (3.1) as the pressure Poisson equation, pressure wave propagation can be observed in the calculation result. The time step size should be small enough to catch the wave propagation.

This type of formulation has been used in the finite volume method as well (Hirt and Nichols, 1980).

Another algorithm has been developed for stronger compressibility (Arai and Koshizuka, 2009). This algorithm is called MPS-AS (MPS for All Speed). The equation of state is written as

$$P = P(\rho, e). \tag{3.6}$$

Here, pressure P is a function of density ρ and the internal energy e. For ideal gas, it can be written as

$$P = (\gamma - 1)\rho e, \tag{3.7}$$

where γ is the heat capacity ratio. For liquid water the following Tait equation is used:

$$\frac{P + B}{P_0 + B} = \left(\frac{\rho}{\rho_0}\right)^N, \tag{3.8}$$

where $P_0 = 1.013 \times 10^5$ Pa, $\rho_0 = 1.000 \times 10^3$ kg/m^3, $B = 2.996 \times 10^8$ Pa, and $N = 7.415$. The time derivative of pressure can be shown as

$$\frac{DP}{Dt} = \frac{\partial P}{\partial e}\frac{De}{Dt} + \frac{\partial P}{\partial \rho}\frac{D\rho}{Dt}. \tag{3.9}$$

The mass, momentum, and energy conservation equations can be written as

$$\frac{1}{\rho}\frac{D\rho}{Dt} + \nabla \cdot \mathbf{u} = 0, \tag{3.10}$$

$$\frac{D\mathbf{u}}{Dt} = -\frac{1}{\rho}\nabla P + \nu\nabla^2\mathbf{u} + \mathbf{g}, \tag{3.11}$$

$$\frac{De}{Dt} = -\frac{P}{\rho}\nabla\cdot\mathbf{u} + \frac{\kappa}{\rho}\nabla^2 T + \Phi, \tag{3.12}$$

where κ is thermal conductivity. The second and third terms in the right hand side of Eq. (3.12) represent heat conduction and viscous heat generation, respectively. Substituting Eqs. (3.10) and (3.12) into Eq. (3.9), we have

$$\frac{DP}{Dt} = -\left(\frac{P}{\rho}\frac{\partial P}{\partial e} + \rho\frac{\partial P}{\partial \rho}\right)\nabla\cdot\mathbf{u} + \left(\frac{\kappa}{\rho}\nabla^2 T + \Phi\right)\frac{\partial P}{\partial e}. \tag{3.13}$$

When the second term in the right hand side of Eq. (3.13) is negligible, the following equation is obtained:

$$\frac{DP}{Dt} = -\left(\frac{P}{\rho}\frac{\partial P}{\partial e} + \rho\frac{\partial P}{\partial \rho}\right)\nabla\cdot\mathbf{u}. \tag{3.14}$$

The procedure in one time step is divided into two processes: prediction and correction as proposed in the CCUP (Cubic-interpolated pseudoparticle for Combined Unified Procedure) method (Yabe and Wang, 1991). In the prediction process the temporary velocities, coordinates, and density are calculated at each particle:

$$\frac{\mathbf{u}_i^* - \mathbf{u}_i^k}{\Delta t} = \nu\langle\nabla^2\mathbf{u}\rangle_i^k + \mathbf{g}, \tag{3.15}$$

$$\frac{\mathbf{r}_i^* - \mathbf{r}_i^k}{\Delta t} = \mathbf{u}_i^k, \tag{3.16}$$

$$\rho_i^* = \frac{\rho_0}{n^0(r_e)}\sum_{j\neq i} w\left(|\mathbf{r}_j^* - \mathbf{r}_i^*|, r_e\right), \tag{3.17}$$

where the effective radius r_e of the weight function is variable to consider relatively large density change. In Eq. (3.15) the momentum conservation equation is explicitly calculated except for the pressure gradient term. Quantity $n^0(r_e)$ is a constant particle number density when the particles are arranged as a uniform square grid with the reference density ρ^0. In the correction process the velocities are implicitly corrected by the pressure gradient term:

$$\frac{\mathbf{u}_i^{k+1} - \mathbf{u}_i^*}{\Delta t} = -\frac{1}{\rho_i^*}\langle\nabla P\rangle_i^{k+1}. \tag{3.18}$$

The new–time pressure is implicitly evaluated by using the Eq. (3.14):

$$\frac{P_i^{k+1} - P_i^k}{\Delta t} = -\left(\frac{P}{\rho}\frac{\partial P}{\partial e} + \rho\frac{\partial P}{\partial \rho}\right)\langle\nabla\cdot\mathbf{u}\rangle_i^{k+1}. \tag{3.19}$$

Divergence of Eq. (3.18) is

$$\frac{\langle \nabla \cdot \mathbf{u} \rangle_i^{k+1} - \langle \nabla \cdot \mathbf{u} \rangle_i^*}{\Delta t} = - \left\langle \nabla \cdot \frac{1}{\rho_i^*} \nabla P \right\rangle_i^{k+1}. \tag{3.20}$$

Substituting Eq. (3.19) into Eq. (3.20) the pressure Poisson equation is obtained as

$$\left\langle \nabla \cdot \frac{1}{\rho_i^*} \nabla P \right\rangle_i^{k+1} = \frac{P_i^{k+1} - P_i^k}{\rho_i^* C_s^2 \Delta t^2} + \frac{\langle \nabla \cdot \mathbf{u} \rangle_i^*}{\Delta t}, \tag{3.21}$$

where

$$C_s^2 = \frac{1}{\rho} \left(\frac{P}{\rho} \frac{\partial P}{\partial e} + \rho \frac{\partial P}{\partial \rho} \right). \tag{3.22}$$

Nonlinear characteristics of the equation of state, for example Eqs. (3.7) or (3.8), can be considered by evaluating Eq. (3.22) at each time step and each particle. The divergence of the temporary velocity vector in Eq. (3.21), the second term of the right hand side, is replaced by the deviation of the particle number density as the same as the MPS method

$$\langle \nabla \cdot \mathbf{u} \rangle_i^* = - \frac{1}{n_i^k} \frac{n_i^* - n_i^k}{\Delta t}. \tag{3.23}$$

Finally, the pressure Poisson equation is written as

$$\left\langle \nabla \cdot \frac{1}{\rho_i^*} \nabla P \right\rangle_i^{k+1} = \frac{P_i^{k+1} - P_i^k}{\rho_i^* C_s^2 \Delta t^2} - \frac{1}{n_i^k} \frac{n_i^* - n_i^k}{\Delta t^2}. \tag{3.24}$$

Eq. (3.24) is basically the same as that of the slight linear compressibility, Eq. (3.4), but nonlinearity of the equation of state is considered.

The new-time coordinates and density are obtained as

$$\frac{\mathbf{r}_i^{k+1} - \mathbf{r}_i^*}{\Delta t} = \mathbf{u}_i^{k+1} - \mathbf{u}_i^*, \tag{3.25}$$

$$\rho_i^{k+1} = \rho_0 \sum_{j \neq i} w \left(\left| \mathbf{r}_j^{k+1} - \mathbf{r}_i^{k+1} \right|, r_e \right). \tag{3.26}$$

An artificial pressure is added as the similar way as Monaghan (1988). When the relative velocity is negative between particles i and j

$$\frac{\left(\mathbf{u}_j - \mathbf{u}_i\right) \cdot \left(\mathbf{r}_j - \mathbf{r}_i\right)}{\left|\mathbf{r}_j - \mathbf{r}_i\right|^2} < 0, \tag{3.27}$$

where two particles i and j are approaching, the following artificial pressure is given

$$\mu_{ij} = r_e \langle \nabla \cdot \mathbf{u} \rangle_{ij}, \tag{3.28}$$

$$P_{a,ij} = \rho_{ij}\left(-\alpha \mu_{ij} C_s + \beta \mu_{ij}^2\right), \tag{3.29}$$

where $\alpha = 1.0$ and $\beta = 2.0$. The velocities of particles i and j are modified as

$$\Delta \mathbf{u}_{ij} = -\Delta t \left\langle \frac{1}{\rho} \nabla P_a \right\rangle_{ij} = -\frac{\Delta t}{n_i} \frac{P_{a,ij}}{\rho_{ij}} \frac{\left(\mathbf{r}_j - \mathbf{r}_i\right)}{\left|\mathbf{r}_j - \mathbf{r}_i\right|^2} w\left(\left|\mathbf{r}_j - \mathbf{r}_i\right|, r_e\right). \tag{3.30}$$

The artificial pressure works as a repulsive force between particles i and j.

In the MPS-AS method the effective radius is a variable to consider large change of density

$$r_{e,i} = r_{e,0} n_i^{-1/d}, \tag{3.31}$$

where d is the spatial dimensions. When the density is larger, the particle number density is also larger. The effective radius is decreased to keep the same value of the particle number density by applying Eq. (3.31).

In the MPS-AS method the following weight function is used:

$$w(r, r_e) = \max\left(\frac{\ln\left(r_e/r\right)}{W_0}, 0\right), \tag{3.32}$$

where W_0 is a normalization function to make the particle number density unity on the square grid arrangement.

To consider variable r_e, the gradient, divergence, and Laplacian models are written as

$$\langle \nabla \phi \rangle_i = \frac{d}{n_i} \sum_{j \neq i} \left[\frac{\phi_j - \phi_i}{\left|\mathbf{r}_j - \mathbf{r}_i\right|^2} \left(\mathbf{r}_j - \mathbf{r}_i\right) w\left(\left|\mathbf{r}_j - \mathbf{r}\right|, r_e\right) \frac{n_i}{n_j} \right], \tag{3.33}$$

$$\langle \nabla \cdot \mathbf{u} \rangle_i = \frac{d}{n_i} \sum_{j \neq i} \left[\frac{(\mathbf{u}_j - \mathbf{u}_i) \cdot (\mathbf{r}_j - \mathbf{r}_i)}{|\mathbf{r}_j - \mathbf{r}_i|^2} w(|\mathbf{r}_j - \mathbf{r}_i|, r_e) \frac{n_i}{n_j} \right], \qquad (3.34)$$

$$\langle \nabla^2 \phi \rangle_i = \frac{2d}{\lambda n_i} \sum_{j \neq i} \left[(\phi_j - \phi_i) w(|\mathbf{r}_j - \mathbf{r}_i|, r_e) \right], \qquad (3.35)$$

where

$$n_i = \sum_{j \neq i} w(|\mathbf{r}_j - \mathbf{r}_i|, r_e), \qquad (3.36)$$

$$\lambda = \frac{1}{n_i} \sum_{j \neq i} |\mathbf{r}_j - \mathbf{r}_i|^2 w(|\mathbf{r}_j - \mathbf{r}_i|, r_e). \qquad (3.37)$$

The differential operators in the governing equations are discretized by using Eqs. (3.33−3.37) in the MPS-AS method.

In this section, two methods are explained as slightly compressible MPS and MPS-AS. Small density change can by analyzed in the slightly compressible MPS method, while large density change can be more accurately analyzed by the MPS-AS method using the equation of state and the variable effective radius.

The MPS-AS method was applied to liquid droplet impingement (LDI) problems in the nuclear power plants (Xiong et al., 2010, 2011, 2012). High-speed steam flow accompanied by small liquid droplets causes LDI on the inner surface of the bent pipes. Repeated impingement of the droplets can lead to the damage on the pipe inner wall. Numerical analysis was carried out in two dimensions. A water droplet that has the diameter of 50 μm and the initial velocity of 200 m/s impinges on a rigid wall. The Tait equation (Eq. 3.8) is used. Fig. 3.1 shows the pressure wave propagation inside the droplet (Xiong et al., 2010). High pressure appears initially at the contact point between the droplet and the wall. The high-pressure front propagates inside the droplet as concentric circles. We can see very thin side jets from the corners.

When the pipe inner surface is wet, which can be simulated by a rigid wall covered with a thin water film, the impingement behavior is changed as shown in Fig. 3.2 (Xiong et al., 2011). In this case a water film of 5 μm thickness is placed on the wall. We can see two pressure waves propagating to the upper and lower directions from the contact point

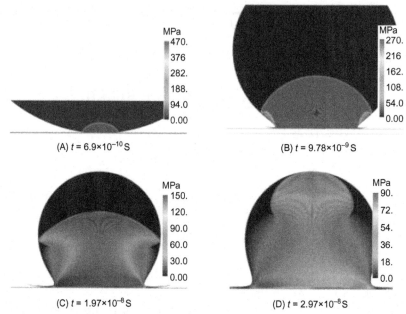

Figure 3.1 Numerical analysis of liquid droplet impingement on a dry rigid wall using MPS-AS (Xiong et al., 2010).

between the droplet and the water film. The pressure wave propagating to the lower direction is reflected by the wall. Then, two pressure waves propagate upward inside the droplet. The Tait equation (Eq. 3.8) leads to a faster propagation speed for higher density and pressure. The second wave is faster than the first wave. Thus the second wave is approaching to the first wave as shown in Fig. 3.2C−E.

3.2 EXPLICIT ALGORITHM USING PSEUDO-COMPRESSIBILITY

In the standard MPS method (Koshizuka and Oka, 1996) the pressure is implicitly solved using the pressure Poisson equation (Eq. 3.1), while the other terms are explicitly solved. This semi−implicit algorithm has been used for the incompressible flow in the finite volume method as well. Solving the pressure Poisson equation is most time-consuming in the semi-implicit algorithm. Shakibaeinia and Jin (2010) replaced the pressure Poisson equation with the equation of state to obtain the pressure explicitly. This makes the algorithm fully explicit. They used the Tait equation

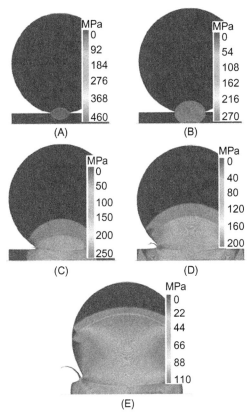

Figure 3.2 Numerical analysis of liquid droplet impingement on a rigid wall covered with a thin liquid film using MPS-AS (Xiong et al., 2011).

(Eq. 3.8) as the equation of state. The sound speed was artificially reduced with keeping the Mach number less than 0.1. They also pointed out that the same time step size can be used without numerical instability, which lead to remarkably faster computation than the standard MPS (Koshizuka and Oka, 1996) using the semi-implicit algorithm. The fully explicit algorithm has been used in Smoothed Particle Hydrodynamics (SPH) and called WCSPH (Weakly Compressible SPH) (Monaghan, 1994).

Here a fully explicit algorithm for MPS described in Yamada et al. (2011) is explained. This method is called E-MPS (Explicit MPS). It has two steps. The first one is to calculate explicitly the Navier-Stokes equations excepting the pressure gradient term, and temporary velocity components and coordinates are obtained. This is the same as the first step of the semi-implicit algorithm. The second one is to calculate explicitly the new-time pressure:

$$P_i^{k+1} = \begin{cases} c^2\left(\rho_i^* - \rho_0\right) & \rho_i^* > \rho_0 \\ 0 & \rho_i^* \leq \rho_0 \end{cases}, \tag{3.38}$$

where c is the sound speed that is given as an artificially small value and ρ_0 is the reference density.

The temporary density is evaluated as

$$\rho_i^* = \frac{\rho_0}{n_0} \sum_{j \neq i} w\left(\left|\mathbf{r}_j^* - \mathbf{r}_i^*\right|\right), \tag{3.39}$$

where the initial particle number density is obtained in the initial particle arrangement:

$$n_0 = \sum_{j \neq i} w\left(\left|\mathbf{r}_j^0 - \mathbf{r}_i^0\right|\right). \tag{3.40}$$

The reference density ρ_0 corresponds to n_0. The pressure is given as a linear function of the deviation from the reference density as shown in the first equation of Eq. (3.38). When the temporary density is lower than the reference density, the pressure is given zero as shown in the second equation of Eq. (3.38). This works as the free surface boundary condition.

The new-time velocity components and coordinates are obtained as the same way as the semi-implicit algorithm using the new-time pressure:

$$\mathbf{u}_i^{k+1} = \mathbf{u}_i^* - \frac{\Delta t}{\rho_0}\left\langle \nabla P^{k+1} \right\rangle_i, \tag{3.41}$$

$$\mathbf{r}_i^{k+1} = \mathbf{r}_i^* + \Delta t\left(\mathbf{u}_i^{k+1} - \mathbf{u}_i^*\right). \tag{3.42}$$

In this E-MPS method the following formulation is used for the discretized pressure gradient in Eq. (3.41):

$$\langle \nabla P \rangle_i = \frac{2d}{n_{0\mathrm{grad}}} \sum_{j \neq i} \left[\frac{P_{ij}}{\left|\mathbf{r}_j - \mathbf{r}_i\right|^2}\left(\mathbf{r}_j - \mathbf{r}_i\right) w_{\mathrm{grad}}\left(\left|\mathbf{r}_j - \mathbf{r}_i\right|\right) \right], \tag{3.43}$$

where

$$P_{ij} = \frac{P_i + P_j}{2}, \tag{3.44}$$

$$n_{0grad} = \sum_{j \neq i} w_{grad}\left(\left|\mathbf{r}_j^0 - \mathbf{r}_i^0\right|\right). \tag{3.45}$$

Here, w_{grad} is the weight function specific to the pressure gradient term. In Yamada et al. (2011), the following weight functions are used:

$$w_{grad}(r) = \begin{cases} \dfrac{r_e}{r} - \dfrac{r}{r_e} & r < r_e \\ 0 & r \geq r_e \end{cases}. \tag{3.46}$$

$$w(r) = \begin{cases} \dfrac{r_e}{r} + \dfrac{r}{r_e} - 2 & r < r_e \\ 0 & r \geq r_e \end{cases}. \tag{3.47}$$

Eq. (3.46) is used for the pressure gradient term and Eq. (3.47) is used for the other terms. Eq. (3.47) is made so that the first-order derivative with respect to r is continuous at r_e.

Eq. (3.43) is different from the normal gradient model in the MPS method. In Eq. (3.43) the right hand side involves $P_j + P_i$, while the normal model involves $P_j - P_i$. Eq. (3.43) can be reduced from the divergence model for the stress tensor in which the pressure is the diagonal components. When $\mathbf{P}_{ij}$ is a stress tensor given between two particles i and j, the divergence model in the MPS method can be written as

$$\langle \nabla \cdot \mathbf{P} \rangle_i = \frac{2d}{n_0} \sum_{j \neq i} \left[\frac{\mathbf{P}_{ij} \cdot (\mathbf{r}_j - \mathbf{r}_i)}{|\mathbf{r}_j - \mathbf{r}_i|^2} w\left(|\mathbf{r}_j - \mathbf{r}_i|\right) \right]. \tag{3.48}$$

When the stress tensor is given as a diagonal matrix as

$$\mathbf{P}_{ij} = \begin{bmatrix} P_{ij} & 0 & 0 \\ 0 & P_{ij} & 0 \\ 0 & 0 & P_{ij} \end{bmatrix}, \tag{3.49}$$

where P_{ij} is the diagonal component. The inner product of the stress tensor and the relative coordinates between two particles i and j are represented by a vector

$$\mathbf{P}_{ij} \cdot (\mathbf{r}_j - \mathbf{r}_i) = P_{ij}(\mathbf{r}_j - \mathbf{r}_i). \tag{3.50}$$

Substituting Eq. (3.50) into the right hand side of Eq. (3.48), we have Eq. (3.43).

For the explicit algorithm based on pseudo-compressibility using the artificially smaller sound speed, the following two problems should be discussed. One is the numerical stability. In the explicit algorithm, both the flow velocity and the sound velocity would determine the numerical stability, while only the flow velocity determines the numerical stability in the semi-implicit algorithm. It is concerned that the time step size would be much reduced in the explicit algorithm, which might lead to more computation time than the semi-implicit algorithm. The other is the accuracy. The artificial compressibility might cause additional errors concerning the density. The density would be substantially enhanced where the pressure increases. These problems were investigated by Oochi et al. (2010) and (2011).

According to Oochi et al. (2010), we can define two Courant numbers C_{flow} and C_{sound} using the maximum flow velocity u_{max} and the sound speed c, respectively:

$$C_{flow} = \frac{u_{max}\Delta t}{l_0}, \tag{3.51}$$

$$C_{sound} = \frac{c\Delta t}{l_0}, \tag{3.52}$$

where Δt and l_0 are the time step size the spacing between two adjacent particles, respectively. A simple two-dimensional shockwave problem was set and the numerical stability was investigated by changing C_{flow} with fixing C_{sound} or changing C_{sound} with fixing C_{flow}. The numerical results showed that the stable solution was obtained with

$$C_{flow} < 0.2, \tag{3.53}$$

$$C_{sound} < 1.0. \tag{3.54}$$

These inequalities are called the Courant conditions. Condition (3.53) is common with the semi-implicit algorithm. It was found that the Courant condition with respect to the flow velocity was severer than that with respect to the sound speed.

When

$$c = 5u_{max}, \tag{3.55}$$

Inequalities (3.53) and (3.54) give us the same condition. Eq. (3.55) can be read as

$$M = \frac{u_{max}}{c} = 0.2, \tag{3.56}$$

which means that the Mach number M is 0.2. In this case the time step size is limited by

$$\Delta t < 0.2 \frac{l_0}{u_{max}} = 1.0 \frac{l_0}{c}. \tag{3.57}$$

If we set the sound speed $0.2u_{max}$ in the explicit algorithm based on the pseudo-compressibility, the time step size is limited by the single Courant condition represented by Inequality (3.57). This means that the same time step size of the semi-implicit algorithm can be used in the explicit algorithm. The computation time is remarkably reduced by using the explicit algorithm since the pressure is explicitly evaluated. In addition, the scaling is $O(N^{1.0})$ in the explicit algorithm where N is the number of particles, while it is approximately $O(N^{1.5})$ in the semi–implicit algorithm that needs to solve the pressure Poisson equation. Thus the speed–up of the E-MPS method is more and more for the larger problems than the standard MPS method.

A concern is degradation of accuracy due to pseudo-compressibility. This arises as change of the density that should be kept constant for the rigorous incompressibility condition. However, it must be noted that the mass is strictly conserved in the particle methods while maintaining the number of the particles. This is different from the finite volume method; mass is changed when the compressibility condition is not satisfied.

The accuracy is verified in Oochi et al. (2011). A stationary water in a rectangular tank was analyzed using the E-MPS method. The bottom pressure was expected to be

$$P = \rho_0 gh, \tag{3.58}$$

where ρ_0, g, and h are the water density, the gravitational acceleration and the water depth, respectively. Using Eq. (3.38), we have

$$\frac{\rho^* - \rho_0}{\rho_0} = \frac{gh}{c^2}. \tag{3.59}$$

If we assume u_{max} as

$$u_{max} = 2\sqrt{gh}, \tag{3.60}$$

Eq. (3.59) can be rewritten as

$$\frac{\rho^* - \rho_0}{\rho_0} = \frac{u_{max}^2}{4c^2} = \frac{M^2}{4}. \tag{3.61}$$

Eq. (3.61) shows that the density change is expected to be 0.01 when $M = 0.2$. The density deviation at the tank bottom is only 1%. If this error is acceptable, the explicit algorithm can be useful. Collapse of a water column was analyzed by the E-MPS method (Oochi et al., 2010). We can see stable behavior of water (Fig. 3.3).

Tsunami run-up analysis has been analyzed by the E-MPS method because the water depth is relatively small and less affected by the density error of the E-MPS method. In addition, a large number of particles are necessary, where the E-MPS method is relatively faster (Murotani et al., 2012, 2014). The explicit algorithm is also preferred in the complicated programming for large-scale parallel computing. On the other hand the explicit algorithm is not advantageous for the fluid dynamics of high viscous fluids. The time step size is rather limited by the viscosity.

Figure 3.3 Calculation result of collapse of a water column using Explicit Moving Particle Semi-implicit (E-MPS) (Oochi et al., 2010).

Particularly, when the viscosity term is implicitly solved, the computation time is mainly governed by the viscosity term and the explicit algorithm for pressure is not very effective.

3.3 SYMPLECTIC SCHEME

Symplectic schemes can be explained as time integration schemes or time marching algorithms, which conserve the mechanical energy. The background is first explained.

There are three basic conservation equations in mechanics: the mass, momentum, and energy conservation equations. In computational mechanics, two of three governing equations, the mass and momentum conservation equations, are usually solved and the other one, the energy conservation equation, is not solved. It is common in both the particle methods and the grid methods. In the particle methods, mass is strictly conserved in the discretized formulation when the number of particles keeping a finite mass is conserved in the simulation. The momentum can be strictly conserved when the spatial discretization of the momentum conservation equation is carefully carried out. However, the mechanical energy conservation seems to be impossible since its governing equation is not solved. The symplectic schemes enable us to conserve the mechanical energy without solving the energy conservation equation. Strictly speaking a value that is very close to the mechanical energy is conserved. The approach is the time integration scheme for Hamiltonian dynamics (Hairer et al., 2000; Leimkuhler and Reich, 2004).

Hamiltonian is the total mechanical energy, which is kinetic energy plus potential energy. We assume that Hamiltonian H is a function of coordinates q_i and momenta p_i of particles i

$$H = H(q_i, p_i).\tag{3.62}$$

Hamilton's canonical equations of motion are written as

$$\frac{d}{dt}q_i = \frac{\partial}{\partial p_i}H(q_i, p_i),\tag{3.63}$$

$$\frac{d}{dt}p_i = -\frac{\partial}{\partial q_i}H(q_i, p_i).\tag{3.64}$$

A linear mapping is called symplectic if the area in the phase space ($p-q$ space) is preserved (Hairer et al., 2000). The time difference scheme is considered as a linear mapping.

The symplectic Euler scheme is first-order symplectic:

$$p_i^{k+1} = p_i^k - \Delta t \frac{\partial}{\partial q_i} H\left(q_i^k, p_i^{k+1}\right), \tag{3.65}$$

$$q_i^{k+1} = q_i^k + \Delta t \frac{\partial}{\partial p_i} H\left(q_i^k, p_i^{k+1}\right). \tag{3.66}$$

Eqs. (3.65) and (3.66) are general description. In the case of fluid dynamics, they can be more specific. Hamiltonian is written as the linear summation of the kinetic energy T that is a function of momenta and the potential energy U that is a function of coordinates

$$H\left(q_i, p_i\right) = T\left(p_i\right) + U\left(q_i\right). \tag{3.67}$$

The momenta and the kinetic energy are specified as

$$p_i = m\mathbf{u}_i, \tag{3.68}$$

$$T\left(p_i\right) = \sum_i \frac{1}{2m} p_i^2 = \sum_i \frac{1}{2} m\mathbf{u}_i^2, \tag{3.69}$$

where m is mass of a particle. The derivative of the right hand side of Eq. (3.69) with respect to p_i is $\mathbf{u}_i$. Then the symplectic Euler scheme can be written as

$$\mathbf{u}_i^{k+1} = \mathbf{u}_i^k - \frac{\Delta t}{m} \nabla_i U\left(\mathbf{r}_i^k\right), \tag{3.70}$$

$$\mathbf{r}_i^{k+1} = \mathbf{r}_i^k + \Delta t \mathbf{u}_i^{k+1}. \tag{3.71}$$

Here the coordinates are replaced to $\mathbf{r}$ to follow the usual description of the particle methods. Roughly speaking, Eqs. (3.70) and (3.71) correspond to the usual time marching algorithm in the particle methods. Firstly, the velocity components are updated explicitly by calculating Eq. (3.70). The second term of the right hand side of Eq. (3.70) stands for the terms in the Navier-Stokes equations at time step k. However, the viscosity term should be excluded here because the mechanical energy dissipates due to the viscosity. Secondly, the coordinates are updated using the updated velocity components as shown in Eq. (3.71). It is also explicitly calculated because the new-time velocity in the right hand side of Eq. (3.71) has already been obtained by solving Eq. (3.70).

The point is the second term of the right hand side of Eq. (3.70). It is the derivative of the potential energy with respect to the coordinates of particle i. We can easily recognize that the gravity term can be written with this formulation if we assume

$$U\left(\mathbf{r}_i^k\right) = \sum_i mgz_i, \tag{3.72}$$

where g is the gravitational acceleration, and the coordinates are denoted by $\mathbf{r}_i = (x_i, y_i, z_i)$. The derivative of Eq. (3.72) with respect to $\mathbf{r}_i$ is obtained as

$$-\nabla_i U\left(\mathbf{r}_i^k\right) = \begin{pmatrix} 0 \\ 0 \\ -mg \end{pmatrix}. \tag{3.73}$$

This represents that the gravitational acceleration g works to the negative z-direction. It is the same as the gravity term in the Navier-Stokes equation.

If the fluid is compressible the internal energy is considered as the potential energy. Here an ideal gas with an adiabatic process is assumed:

$$P = C\rho^\gamma, \tag{3.74}$$

$$P = (\gamma - 1)\rho U, \tag{3.75}$$

where γ is the heat capacity ratio and C is a constant. Combining Eqs. (3.74) and (3.75), we have

$$U = C\frac{1}{\gamma - 1}\rho^{\gamma-1}. \tag{3.76}$$

When the total internal energy is the summation of the internal energy of all particles

$$U = \sum_j m_j U_j, \tag{3.77}$$

the derivative of Eq. (3.72) with respect to $\mathbf{r}_i$ is obtained as

$$
\begin{aligned}
-\nabla_i U &= -\frac{\partial}{\partial \mathbf{r}_i} \sum_j m_j U_j \\
&= -\frac{\partial}{\partial \mathbf{r}_i} \sum_j m_j C \frac{1}{\gamma - 1} \rho_j^{\gamma - 1} \\
&= -\sum_j m_j C \rho_j^{\gamma - 2} \frac{\partial \rho_j}{\partial \mathbf{r}_i} \\
&= -m_i C \rho_i^{\gamma - 2} \frac{\partial \rho_i}{\partial \mathbf{r}_i} - \sum_{j \neq i} m_j C \rho_j^{\gamma - 2} \frac{\partial \rho_j}{\partial \mathbf{r}_i}
\end{aligned} \tag{3.78}
$$

If the density is evaluated by SPH

$$
\rho_i = \sum_j m_j w\left(\left|\mathbf{r}_j - \mathbf{r}_i\right|\right), \tag{3.79}
$$

where $w(r)$ is the kernel function, Eq. (3.78) becomes

$$
\begin{aligned}
-\nabla_i U &= -m_i C \rho_i^{\gamma - 2} \sum_{j \neq i} m_j \nabla_i w_{ij} - \sum_{j \neq i} m_j C \rho_j^{\gamma - 2} m_i \nabla_i w_{ij} \\
&= -m_i \sum_{j \neq i} \left[m_j C \rho_i^{\gamma - 2} \nabla_i w_{ij} + m_j C \rho_j^{\gamma - 2} \nabla_i w_{ij} \right] \\
&= -m_i \sum_{j \neq i} m_j \left[\frac{p_i}{\rho_i^2} + \frac{p_j}{\rho_j^2} \right] \nabla_i w_{ij}
\end{aligned} \tag{3.80}
$$

Since the derivative of the kernel at $\mathbf{r}_i$ is 0, we can write

$$
-\nabla_i U = -m_i \sum_j m_j \left[\frac{p_i}{\rho_i^2} + \frac{p_j}{\rho_j^2} \right] \nabla_i w_{ij}. \tag{3.81}
$$

The right hand side is the same formulation of the pressure gradient term in SPH (Monaghan and Price, 2001; Inutsuka, 2002).

It is shown that the distinctive formulation of the pressure gradient term of SPH can be derived from the Hamilton's canonical equations. Thus SPH is symplectic for compressible fluid dynamics.

The Störmer-Verlet scheme (also called the leap-frog scheme) is known as a second-order explicit symplectic scheme:

$$
\mathbf{u}_i^{k+1/2} = \mathbf{u}_i^k - \frac{\Delta t}{2m} \nabla_i U\left(\mathbf{r}_i^k\right), \tag{3.82}
$$

$$
\mathbf{r}_i^{k+1} = \mathbf{r}_i^k + \Delta t \mathbf{u}_i^{k+1/2}. \tag{3.83}
$$

$$\mathbf{u}_i^{k+1} = \mathbf{u}_i^{k+1/2} - \frac{\Delta t}{2m} \nabla_i U\left(\mathbf{r}_i^{k+1}\right), \tag{3.84}$$

Eqs. (3.82) and (3.84) can be added to

$$\mathbf{u}_i^{k+1/2} = \mathbf{u}_i^{k-1/2} - \frac{\Delta t}{m} \nabla_i U\left(\mathbf{r}_i^k\right). \tag{3.85}$$

The second equation of Hamilton's canonical equations

$$\frac{d}{dt}\mathbf{p}_i = -\frac{\partial}{\partial \mathbf{q}_i} H\left(\mathbf{q}_i, \mathbf{p}_i\right), \tag{3.64}$$

means that the force acting on particle i is obtained as the negative value of the spatial derivative of Hamiltonian at particle i. $\frac{\partial H}{\partial \mathbf{q}_i}$ is the infinitesimal energy increase ∂H divided by the infinitesimal displacement of particle $\partial \mathbf{q}_i$. Thus this expresses the principle of virtual work.

If we make a formulation of total mechanical energy, Hamiltonian, as a differential equation and it is discretized using the particles, mechanical energy conservation is achieved by applying the first-order or second-order symplectic scheme to the discretized Hamiltonian as the time integration scheme. The point is that the discretized motion equation is reduced from the discretized Hamilton's canonical equations.

Next, we consider extension of the previous symplectic schemes to incompressible fluids, which can be regarded as a Hamiltonian system with a constraint condition of incompressibility (Szmidt, 1998). The constraint condition can be written as

$$J - 1 = 0, \tag{3.86}$$

where J is the Jacobian of the transformation. Physical meaning of the Jacobian is the volume change. Eq. (3.86) requires that the volume is kept constant.

Lagrangian can be described as

$$L\left[\mathbf{q}_t, \dot{\mathbf{q}}_t, \lambda_t\right] = \int_{\Omega_0} \left[\frac{1}{2}\rho_0 \left|\dot{\mathbf{q}}_t(\mathbf{a})\right|^2 - \rho_0 U(\mathbf{q}_t(\mathbf{a})) - \lambda_t(\mathbf{a})(1 - J_t(\mathbf{a}))\right] d\mathbf{a}, \tag{3.87}$$

where $\mathbf{a}$ is a label of the fluid particle or the Lagrangian coordinate in the reference configuration Ω_0, ρ_0 is the density, and λ_t is the Lagrange multiplier (Suzuki et al., 2007). The Jacobian J_t is expressed as

$$J_t = \det\left(\frac{\partial \mathbf{q}_t}{\partial \mathbf{a}}\right). \tag{3.88}$$

The first, second, and third terms in the integration of the right hand side of Eq. (3.87) represent the kinetic energy, the potential energy, and the constraint condition, respectively. The constraint condition, incompressibility, is included in the Lagrangian with a Lagrange multiplier. Physically the Lagrange multiplier represents pressure. In incompressible fluids, pressure does not contribute to the potential energy any more.

In Suzuki et al. (2007) the discretized Lagrangian with N particles for incompressible fluids is given by

$$L[\boldsymbol{q}_1, ..., \boldsymbol{q}_N, \dot{\boldsymbol{q}}_1, ..., \dot{\boldsymbol{q}}_N, \lambda_1, ..., \lambda_N]$$

$$= \sum_{j=1}^{N} \left[\frac{1}{2} m_j \left| \dot{\boldsymbol{q}}_j(t) \right|^2 - m_j U(\boldsymbol{q}_j(t)) - \lambda_j(t) \left(\left| \Delta \mathbf{a}_j \right| - \frac{m_j}{\rho\left(\boldsymbol{q}_j(t), t\right)} \right) \right] \cdot$$

$$(3.89)$$

Here $\left| \Delta \mathbf{a}_j \right|$ and m_j are the volume and the mass of particle j, respectively. The mass and the density of particle j are obtained by

$$m_j = \rho_0 \left| \Delta \mathbf{a}_j \right|, \tag{3.90}$$

$$\rho\left(\boldsymbol{q}_j(t), t\right) = \sum_{k=1}^{N} m_k w_{jk}\left(\left| \mathbf{q}_k(t) - \mathbf{q}_j(t) \right| \right). \tag{3.91}$$

Eq. (3.89) can be rewritten as

$$L[\boldsymbol{q}_1, ..., \boldsymbol{q}_N, \dot{\boldsymbol{q}}_1, ..., \dot{\boldsymbol{q}}_N, \lambda_1, ..., \lambda_N]$$

$$= \sum_{j=1}^{N} \left[\frac{1}{2} m_j \left| \dot{\mathbf{q}}_j(t) \right|^2 - m_j U(\boldsymbol{q}_j(t)) - \lambda_j(t) g_j(\boldsymbol{q}_1(t), ..., \boldsymbol{q}_N(t)) \right], \tag{3.92}$$

where

$$g_j(\boldsymbol{q}_1(t), ..., \boldsymbol{q}_N(t)) = \left| \Delta \mathbf{a}_j \right| - \frac{m_j}{\sum\limits_{k=1}^{N} m_k w_{jk}\left(\left| \mathbf{q}_k(t) - \mathbf{q}_j(t) \right| \right)}, \tag{3.93}$$

is the discretized incompressibility condition.

In Eq. (3.91), density is represented by the summation of the weight function. This is the same as SPH and MPS. The incompressibility condition (Eq. 3.93) is that the density is constant, which is the same as MPS.

It is known that the following RATTLE scheme is symplectic for the Hamiltonian system with constraints (Hairer et al., 2000; Leimkuhler and Reich, 2004).

$$p_i^{k+1/2} = p_i^k - \frac{\Delta t}{2}\left[\sum_{j=1}^{N} \lambda_j^{k+1/2} \frac{\partial g_j}{\partial q_i}(q_1^k, ..., q_N^k) + m_i U(q_i^k)\right], \qquad (3.94)$$

$$q_i^{k+1} = q_i^k + \Delta t \frac{p_i^{k+1/2}}{m_i}, \qquad (3.95)$$

$$g_i(q_1^{k+1}, ..., q_N^{k+1}) = 0, \qquad (3.96)$$

$$p_i^{k+1} = p_i^{k+1/2} - \frac{\Delta t}{2}\left[\sum_{j=1}^{N} \lambda_j^{k+1} \frac{\partial g_j}{\partial q_i}(q_1^{k+1}, ..., q_N^{k+1}) + m_i U(q_i^{k+1})\right], \qquad (3.97)$$

$$\sum_{j=1}^{N} \frac{\partial g_i}{\partial q_j}(q_1^{k+1}, ..., q_N^{k+1}) \cdot \frac{p_j^{k+1}}{m_j} = 0. \qquad (3.98)$$

The process to solve the previous equation system is as follows. First, Eq. (3.94) is substituted to $p_i^{k+1/2}$ in Eq. (3.95), Eq. (3.95) is substituted to q_i^{k+1} in Eq. (3.96), and then unknowns in Eq. (3.96) are changed to $\lambda_i^{k+1/2}$. This modified Eq. (3.96) is nonlinear and solved by the Newton-Raphson iteration method. Then we have $\lambda_i^{k+1/2}$. Substituting $\lambda_i^{k+1/2}$ into Eq. (3.94), we have $p_i^{k+1/2}$. Substituting $p_i^{k+1/2}$ into Eq. (3.95), we have q_i^{k+1}. Next, Eq. (3.97) is substituted to p_j^{k+1} in Eq. (3.98), and then unknowns are changed to λ_i^{k+1}. The modified Eq. (3.98) is linear. Solving the simultaneous linear equations, we obtain λ_i^{k+1}. p_i^{k+1} is calculated by substituting λ_i^{k+1} into Eq. (3.97).

Eq. (3.96) written with the Lagrange multiplier $\lambda_i^{k+1/2}$ as unknowns is like a pressure Poisson equation of which the source term is represented by the density being constant, while Eq. (3.98) written with λ_i^{k+1} is also another pressure Poisson equation of which source term is represented by divergence of the velocity being zero. The first pressure equation is the direct representation of the constraint condition. The second pressure equation requires that the motion obeys the derivative of the constraint condition. If we consider the ball motion on a slope. The slope position is the constraint condition where the ball has to stay on the slope. The ball velocity should be parallel to the slope, which is the derivative of the constraint condition. The divergence of the velocity being zero, which is called the divergence-free condition, is the derivative of the incompressibility condition.

Water wave propagation was analyzed using the Hamiltonian MPS (HMPS) method (Suzuki et al., 2007) in two dimensions. Fig. 3.4 illustrates the initial condition. The particles were arranged on a grid where the spacing was $\Delta r = 0.01$ m. The calculation result of the total energy is

Figure 3.4 Initial condition of water wave propagation problem, $\Delta r = 0.01$ m (Suzuki et al., 2007).

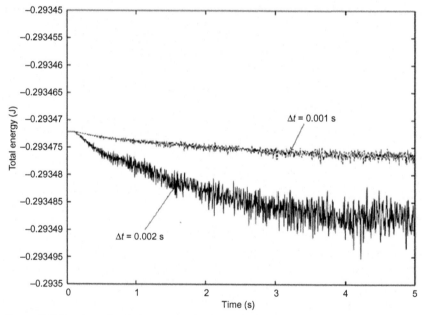

Figure 3.5 Total mechanical energy using Hamiltonian Moving Particle Semi-implicit (HMPS) (Suzuki et al., 2007).

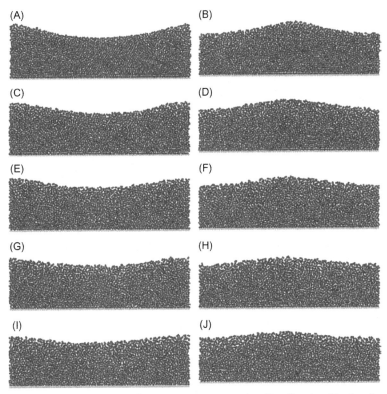

Figure 3.6 Calculation result of wave propagation using Hamiltonian Moving Particle Semi-implicit (HMPS): (A) 0.4 s, (B) 0.8 s, (C) 1.2 s, (D) 1.6 s, (E) 2.0 s, (F) 2.4 s, (G) 2.8 s, (H) 3.2 s, (I) 3.6 s, (J) 4.0 s (Suzuki et al., 2007).

shown in Fig. 3.5. The energy decrease was 0.000015 J for 5 s when $\Delta t = 0.002$ s, and it was 0.000005 J when $\Delta t = 0.001$ s. The decreases were only 0.005% and 0.0017% of the initial total mechanical energy of -0.29 J. We can see very good conservation of the mechanical energy for a long time. However, the wave amplitude decreased so much as shown in Fig. 3.6. The total mechanical energy conservation does not mean good accuracy for wave propagation. The initial energy of the wave was kept accurately though it substantially changed to fluctuating motion.

The same way was also examined for the elastic solid using the MPS method (Suzuki and Koshizuka, 2008; Kondo et al., 2010a). Hamilton's canonical equations were formulated with the symplectic schemes. The calculation examples showed that the total mechanical energy conservation was excellent. An incompressible nonlinear elastic material obeying the Mooney-Rivlin constitutive law was analyzed using the RATTLE algorithm (Suzuki and Koshizuka, 2008). The total energy conservation was very good for a long time as expected.

3.4 ARBITRARY LAGRANGIAN-EULERIAN

The particle methods are usually based on the Lagrangian description, where the variables located at the particle move with the fluid. The convection terms in the Navier–Stokes equations are not necessary to evaluate. On the other hand the grid methods are mainly based on the Eulerian description, where the variables located at the nodes are fixed in space. The convection terms are necessary to evaluate. Here the particles in the particle methods and the nodes in the grid methods have the same meaning that the variables, velocity components, pressure, temperature, etc. are located. If particles or nodes are moved arbitrarily the description is called arbitrary Lagrangian-Eulerian (ALE). In the ALE description the convection terms are also necessary to evaluate.

In the ALE description, we can consider three particle positions in each time step as shown in Fig. 3.7. The position of particle i at time step k is denoted by $\mathbf{r}_i^k$. The velocity is updated from $\mathbf{u}_i^k$ to $\mathbf{u}_i^L$ after the explicit and implicit steps of the MPS algorithm in this time step. The particle moves to

$$\mathbf{r}_i^L = \mathbf{r}_i^k + \Delta t \mathbf{u}_i^L, \tag{3.99}$$

in the fully Lagrangian description. All variables f are moved with the particle keeping the same values, i.e.,

$$f^{k+1}(\mathbf{r}_i^L) = f^k(\mathbf{r}_i^k). \tag{3.100}$$

The new-time position of particle i is then arbitrarily determined as $\mathbf{r}_i^{k+1}$ with a velocity of the particle $\mathbf{u}_i^c$

$$\mathbf{r}_i^{k+1} = \mathbf{r}_i^k + \Delta t \mathbf{u}_i^c. \tag{3.101}$$

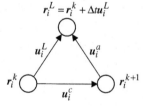

Figure 3.7 Movement of a particle in Arbitrary Lagrangian-Eulerian (ALE).

If the new position is the same as that of fully Lagrangian motion

$$\mathbf{r}_i^{k+1} = \mathbf{r}_i^L, \tag{3.102}$$

$$\mathbf{u}_i^c = \mathbf{u}_i^L, \tag{3.103}$$

This is the standard MPS method based on the Lagrangian description. If the new position is the same as the old position at time step k

$$\mathbf{r}_i^{k+1} = \mathbf{r}_i^k, \tag{3.104}$$

$$\mathbf{u}_i^c = 0. \tag{3.105}$$

This is the fully Eulerian description. If the new position is arbitrarily given, we need to evaluate the change of variables as

$$f^{k+1}(\mathbf{r}_i^{k+1}) = f^{k+1}(\mathbf{r}_i^L - \Delta t \mathbf{u}_i^a), \tag{3.106}$$

$$\mathbf{u}_i^a = \mathbf{u}_i^L - \mathbf{u}_i^c. \tag{3.107}$$

To calculate Eq. (3.106), we need to obtain the spatial distribution of f at time step $k+1$. This is equivalent to the calculation of the convection term.

In general the particle movement in one time step is smaller than that of the particle distance so as to satisfy the Courant condition. Thus the calculation of Eq. (3.106) is an interpolation of the variables at the particles.

Yoon et al. (1999a) proposed a method to calculate the interpolation equivalent to the convection in the ALE description. This method was named "meshless advection using flow-directional local-grid (MAFL)" and combined with MPS. First, a one-dimensional local grid through particle i is generated as shown in Fig. 3.8 in the flow direction of $\mathbf{u}_i^a$. Temporary grid points are given; in Fig. 3.8, three additional grid points $i+1$, $i-1$, and $i-2$ are set with the spacing of Δr. Values at the temporary grid points are evaluated by weighted averages of their neighboring areas Ω_{i+1}, Ω_{i-1}, and Ω_{i-2}, respectively. The neighboring area is limited by a circle of radius r_e and lines vertical to the grid.

The new-time value at particle i is calculated, for instance, by the QUICK scheme (Leonard, 1979) as

$$f_i^{k+1} = f_i^L - \frac{\Delta t}{\Delta r}|\mathbf{u}_i^a| \left(\frac{3}{8}f_{i+1} + \frac{3}{8}f_i - \frac{7}{8}f_{i-1} + \frac{1}{8}f_{i-2} \right). \tag{3.108}$$

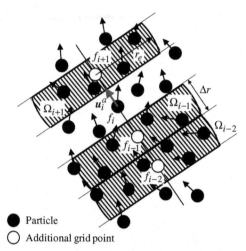

Figure 3.8 Flow-directional local grid in meshless advection using flow-directional local-grid (MAFL).

The calculation of the convection term with the flow velocity of $\mathbf{u}_i^a$ is equivalent to evaluate the interpolated value at the upwind position of $\mathbf{r}_i^L - \Delta t \mathbf{u}_i^a$ on the flow-directional local grid. This concept is called "semi-Lagrangian" and widely used in the finite difference method.

In MPS-MAFL the temporary grid is generated for each particle at each time step, so that the global mesh is not necessary and mesh tangling does not occur. Since the temporary grid is one-dimensional, the calculation is not complicated even in three dimensions. Furthermore the knowledge of the finite difference schemes can be applied to the temporary grid. Thus MPS-MAFL is practical in the ALE description without losing the advantages of the particle methods.

In MPS-MAFL the source term of the pressure Poisson equation is represented by the divergence of the velocity because the particles are arbitrarily located and the particle number density is not proportional to the fluid density:

$$\langle \nabla \cdot \mathbf{u} \rangle_i = \frac{d}{n_i} \sum_{j \neq i} \left[\frac{(\mathbf{u}_j - \mathbf{u}_i) \cdot (\mathbf{r}_j - \mathbf{r}_i)}{|\mathbf{r}_j - \mathbf{r}_i|^2} w(|\mathbf{r}_j - \mathbf{r}_i|) \right], \tag{3.109}$$

$$\langle \nabla^2 P^{k+1} \rangle_i = \frac{\rho}{\Delta t} \langle \nabla \cdot \mathbf{u}^* \rangle_i, \tag{3.110}$$

$$\frac{u_i^L - u_i^*}{\Delta t} = -\frac{1}{\rho}\left\langle \nabla P^{k+1}\right\rangle_i. \tag{3.111}$$

A calculation example of flow-induced sloshing is shown in Fig. 3.9. Sloshing in a rectangular tank is excited by the flow entering from the left side and going out from the bottom (Yoon et al., 1999a). The inlet and outlet boundaries are calculated by the fixed particles (Eulerian),

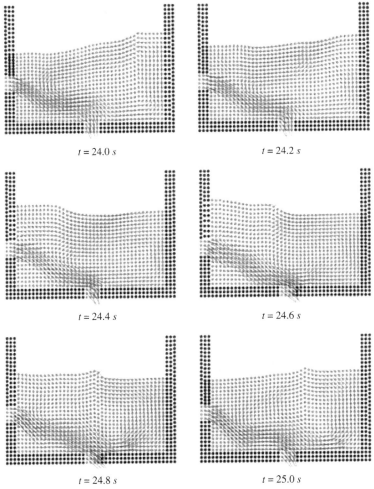

$t = 24.0\ s$ $t = 24.2\ s$

$t = 24.4\ s$ $t = 24.6\ s$

$t = 24.8\ s$ $t = 25.0\ s$

Figure 3.9 Calculation result of flow-induced sloshing in a rectangular tank using Moving Particle Semi-implicit-meshless advection using flow-directional local-grid (MPS-MAFL) (Yoon et al., 1999a).

while the oscillating free surface is calculated by the moving particles (almost Lagrangian). The Eulerian treatment is simpler on the inlet and outlet boundaries than the Lagrangian treatment where the new particles are generated in the inlet and the outflowing particles are removed. However, the Lagrangian treatment is preferred on the moving free surface. The ALE treatment is particularly useful for this type of problems. The excitation of sloshing is affected by the water level and the inflow velocity. The calculated excitation condition showed good agreement with that of the experiment.

One cycle of the bubble growth and departure in subcooled nucleate boiling was simulated using MPS-MAFL in two dimensions (Yoon et al., 2001). The result is shown in Fig. 3.10. A small bubble was initially located on the heated wall of which temperature was 110°C. The pressure was atmospheric and the bulk water temperature was 96°C. Steam in the bubble was assumed to be uniform and saturated. No particles were used in steam.

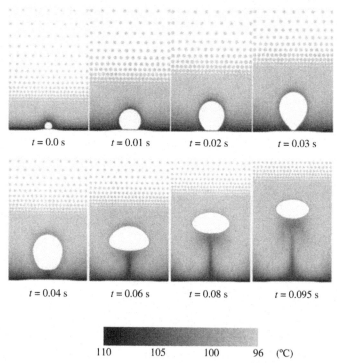

Figure 3.10 Bubble growth and departure in nucleate boiling: heated wall temperature 110°C, saturation temperature 100°C, and bulk temperature 96°C (Yoon et al., 2001).

The initial small bubble grew rapidly due to mass transfer from the thin temperature boundary layer near the hot wall on the bottom. The bubble detached from the bottom wall at around 0.03 s due to buoyancy force and rose up. Hot water rose up as well to follow the bubble motion. The bubble shrank due to condensation in the subcooled bulk water. The bubble growth and consequent heat transfer agreed well with the experiment.

In this problem, small particles were placed in the thin boundary layers near the hot wall and the bubble surface, while larger particles were located in the bulk water where high spatial resolution is not necessary. The small particles stayed in the boundary layers using the ALE treatment. On the other hand it would not be possible in the fully Lagrangian treatment where the small particles move with their own velocities and some of them might flow out of the boundary layers. We can see that the number of particles increases when the bubble grows and rises up. More small particles are necessary for the boundary layer around the bubble. MPS-MAFL is applicable to both moving particles and new particles because the calculation procedure is the same for them. We need to interpolate a value at an arbitrary position from those at the existing particle positions. We do not need to care that the particle is moved or newly generated.

MPS-MAFL has been used in Yoon et al. (1999a, b, 2001), Koshizuka et al. (2000), and Heo et al. (2002).

Hu et al. (2017) proposed a new approach for the ALE particle method. The convection scheme is evaluated using LSMPS (Least Squares MPS) (Tamai et al., 2013; Tamai and Koshizuka, 2014) applied to the upwind half of the neighboring particles (Fig. 3.11). This can be expressed by the upwind weight function

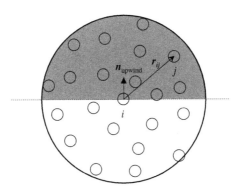

Figure 3.11 Upwind weight function for Arbitrary Lagrangian-Eulerian (ALE) (Hu et al., 2017).

$$w_{\text{upwind}}(r) = \begin{cases} w(|\mathbf{r}_{ij}|) & \mathbf{r}_{ij} \cdot \mathbf{n}_{\text{upwind},i} > 0 \\ 0 & \text{otherwise} \end{cases}, \tag{3.112}$$

$$\mathbf{r}_{ij} = \mathbf{r}_j - \mathbf{r}_i, \tag{3.113}$$

$$n_{\text{upwind},i} = \mathbf{u}_i^a / |\mathbf{u}_i^a|, \tag{3.114}$$

where $\mathbf{r}_i$ and $\mathbf{r}_j$ are position vectors of particles i and j, respectively. The proposed ALE particle method showed that accuracy was much improved than MPS-MAFL in a lid-driven cavity flow problem.

A particle shifting technique (Xu et al., 2009) was proposed for rearrangement of the particles to avoid clustering in Incompressible Smoothed Particle Hydrodynamics (ISPH) when the velocity divergence is employed as the source term in the pressure Poisson equation. After the particle shifting the variables at the particles are updated by using the first-order derivative and the distance vector, which is equivalent to evaluating the convection terms in the ALE approach.

3.5 RIGID BODY MODEL

A rigid body is a solid that has a finite mass and a finite volume. Deformation is negligible and the shape is not changed. A rigid body has six degrees of freedom in three-dimensional space: three translational velocity components and three angular velocity components. In the particle method a rigid body is represented by a single particle or connection of multiple particles possessing the same mass and the same volume.

In two-dimensional space a rigid body has three degrees of freedom: two velocity components (u, v) or $(u, v, 0)$ and one angular velocity ω or $\boldsymbol{\omega} = (0, 0, \omega)$. A rigid body is represented by N solid particles of coordinates $\mathbf{r}_i$ and velocities $\mathbf{u}_i$ (Fig. 3.12). The coordinates of the center of gravity $\mathbf{r}_g$ are obtained by

$$\mathbf{r}_g = \frac{1}{N} \sum_{i=1}^{N} \mathbf{r}_i. \tag{3.115}$$

The relative coordinates of the solid particles with respect to the center of gravity are denoted by:

$$\boldsymbol{q}_i = \mathbf{r}_i - \mathbf{r}_g. \tag{3.116}$$

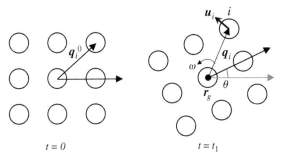

Figure 3.12 A rigid body represented by solid particles in two-dimensional space.

The translational velocity and the angular velocity of the center of gravity are calculated by

$$\mathbf{u}_g = \frac{1}{N} \sum_{i=1}^{N} \mathbf{u}_i, \tag{3.117}$$

$$\boldsymbol{\omega} = \frac{1}{I} \sum_{i=1}^{N} \mathbf{q}_i \times \mathbf{u}_i, \tag{3.118}$$

where

$$I = \sum_{i=1}^{N} |\mathbf{q}_i|^2. \tag{3.119}$$

Here the coordinates and velocities have only two components in x and y directions and the angular velocity has one component in z-direction. When the rigid body keeps its shape, the velocities of the solid particles belonging to the rigid body can be reproduced by

$$\mathbf{u}_i = \mathbf{u}_g + \mathbf{q}_i \times \boldsymbol{\omega}. \tag{3.120}$$

The coordinates of the solid particles can be reproduced by

$$\mathbf{q}_i = \mathbf{R}(\theta)\mathbf{q}_i^0, \tag{3.121}$$

$$\frac{d\theta}{dt} = \omega, \tag{3.122}$$

where $\mathbf{q}_i^0$ is the initial relative coordinates and $\mathbf{R}(\theta)$ is the rotation matrix determined by the angle θ from the initial position. Substituting Eq. (3.121) into Eq. (3.116), we have $\mathbf{r}_i$.

Koshizuka et al. (1998) calculated interaction between a fluid and a rigid body using the following algorithm. First, both fluid and solid particles were calculated and moved as an incompressible fluid together for one time step using the MPS method. After the explicit and implicit procedures the shape of the rigid body was not kept. The new–time coordinates of the center of gravity of the rigid body was obtained by Eq. (3.115). The translational velocity and angular velocity of the center of gravity of the rigid body were evaluated by Eqs. (3.117) and (3.118), respectively. The new-time angle is updated using Eq. (3.122). The new-time velocities and coordinates of the solid particles were given by Eqs. (3.120), (3.121), and (3.116).

The algorithm mentioned previously is simple and useful since all the fluid and solid particles are first calculated together and then only the solid particles are modified to maintain the initial shape of the rigid body while keeping the translational and angular momenta. This algorithm is justified by the divergence theorem of Gauss

$$\iiint_V \nabla P d\boldsymbol{v} = \oiint_S P d\boldsymbol{s}, \qquad (3.123)$$

$$\iiint_V \mathbf{r} \times \nabla P d\boldsymbol{v} = \oiint_S \mathbf{r} \times P d\boldsymbol{s}. \qquad (3.124)$$

The equations mentioned previously say that the surface integration of pressure can be expressed by the volume integration of pressure gradient. The surface integration of pressure is usually used for the interaction between the solid and the fluid. The volume integration of pressure gradient in the Navier-Stokes equations can be used as well for the interaction (Shibata et al., 2012). Calculation of the pressure gradient term for this type of algorithm is also used in Computer Graphics (Carlson et al., 2004).

The governing equations of rigid bodies are regarded as the Navier-Stokes equations with the constraint conditions that the initial shapes of the rigid bodies are kept.

Modeling of a three-dimensional rigid body was explained by Shibata et al. (2012). A rigid body represented by N solid particles has six degrees of freedom: translational velocity components $\mathbf{u} = (u, v, w)$ and angular velocity components $\boldsymbol{\omega} = (\omega_x, \omega_y, \omega_z)$. Corresponding coordinates for the translational velocity is the coordinates of the center of gravity $\mathbf{r}_g = (x_g, y_g, z_g)$.

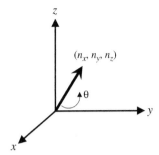

Figure 3.13 Quaternion for the angle system in three-dimensional space.

Since the angular system in three-dimensional space is nonlinear, we need a special treatment. A quaternion was employed for the angular system $q = (a, b, c, d)$

$$a = n_x \sin\frac{\theta}{2}, \quad b = n_y \sin\frac{\theta}{2}, \quad c = n_z \sin\frac{\theta}{2}, \quad d = \cos\frac{\theta}{2}, \tag{3.125}$$

where the unit vector $\mathbf{n} = (n_x, n_y, n_z)$ and its rotating angle θ are defined as shown in Fig. 3.13. The initial relative coordinates of the particles are transformed to the present relative coordinate by rotating θ with respect to the axis $\mathbf{n}$. A quaternion contains four variables, while the degrees of freedom are three. One constraint condition is added to the quaternion.

$$1 = a^2 + b^2 + c^2 + d^2. \tag{3.126}$$

The quaternion has no singularities as an angle system in three-dimensional space.

We can use the rotation matrix, such as Eq. (3.121), as the angular system without singularities. But it has nine components in three-dimensional space and six constraint conditions are additionally required. A quaternion has four variables with one constraint condition. The relationship between the rotation matrix in three dimensions and the quaternion is as follows:

$$\mathbf{R} = \begin{bmatrix} 1 - 2b^2 - 2c^2 & 2ab - 2dc & 2ac + 2db \\ 2ab + 2dc & 1 - 2a^2 - 2c^2 & 2bc - 2da \\ 2ac - 2db & 2bc + 2da & 1 - 2a^2 - 2b^2 \end{bmatrix}. \tag{3.127}$$

In Computer Graphics the quaternion has widely been used for the rigid body simulation.

The moment of inertia is a tensor in three-dimensional space. For a rigid body the principal moments of inertia, where the tensor of inertia is diagonal, can be obtained around the principal axes

$$
\boldsymbol{I} = \begin{bmatrix} I_1 & 0 & 0 \\ 0 & I_2 & 0 \\ 0 & 0 & I_3 \end{bmatrix}.
\tag{3.128}
$$

If the principal axes are the same as those of the Cartesian coordinate system in the initial condition, the principal moments of inertia are the same as the initial one. The present moments of inertia are calculated by using the rotation matrix as

$$
\boldsymbol{I} = \boldsymbol{R}\boldsymbol{I}_0\boldsymbol{R}^T,
\tag{3.129}
$$

where $\boldsymbol{I}$ and $\boldsymbol{I}_0$ are the present and initial moments of inertia and superscript T stands for the transposed matrix. Eq. (3.129) means that the present moments of inertia of the rigid body are changing by its rotating motion.

We can solve the rotating motion using the principal axes. Euler's equations

$$
\begin{cases}
M_1 = I_1 \dfrac{d\omega_1}{dt} - (I_2 - I_3)\omega_2\omega_3 \\[2ex]
M_2 = I_2 \dfrac{d\omega_2}{dt} - (I_3 - I_1)\omega_3\omega_1 \,, \\[2ex]
M_3 = I_3 \dfrac{d\omega_3}{dt} - (I_1 - I_2)\omega_1\omega_2
\end{cases}
\tag{3.130}
$$

are solved around the principal axes where the moments of inertia are not changed as $\boldsymbol{I}_0$. Here, $(\omega_1, \omega_2, \omega_3)$ and (M_1, M_2, M_3) are angular velocities and torques around the principal axes, respectively.

An example of interaction between a rigid body and a fluid in three-dimensional space is depicted in Fig. 3.14 (Shibata et al., 2012). An advancing model ship was represented by a rigid body, which interacted with regular waves. The ship had 3 m length and heave and pitch motions were free. The waves were 3 m length and 0.108 m height. In Fig. 3.14, we can see that water goes up on the ship deck as the ship bow moves down into the wave crest by a pitching motion.

Another example is shown in Fig. 3.15. All buildings and houses were released as floating rigid bodies in a large-scale tsunami run–up analysis

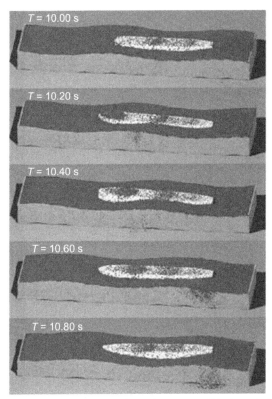

Figure 3.14 Analysis of ship motion and regular waves: Ship length 3 m, wavelength 3 m, and wave height 0.108 m (Shibata et al., 2012).

Figure 3.15 Tsunami run-up analysis in coastal city with virtual release of buildings and houses as rigid bodies (Murotani et al., 2014).

(Murotani et al., 2014). Good performance of parallel computing was maintained when many rigid bodies were introduced.

Tsukamoto et al. (2011) analyzed sloshing with a floating rigid body connected by springs to the wall. Damping of waves was successfully calculated by the effect of fluid-rigid body interaction.

3.6 STRUCTURAL ANALYSIS

To conduct fluid—structure interaction, contact between particles can easily be detected just by checking the distance between them. Therefore, when the structure analysis can also be done using particles, fluid—structure interaction problem can totally be treated in a particle system. Since the elastic solid analysis is the most basic structure analysis, several particle models for elastic analysis have been proposed.

An elastic solid model based on the MPS method is explained. An elastic solid recovers its shape after deformation such as a spring recovers its length after stretching. Therefore when the particles are connected with springs they behave like an elastic solid. Song et al. (2005) developed an elastic solid model based on this idea. The governing equation for the elastic solid can be rewritten as

$$
\begin{aligned}
\rho \frac{d\mathbf{v}}{dt} &= \nabla \cdot (\lambda \mathrm{tr}(\boldsymbol{\varepsilon})\mathbf{I} + 2\mu\boldsymbol{\varepsilon}) \\
&= \lambda \nabla \mathrm{tr}(\boldsymbol{\varepsilon}) + 2\mu \nabla^2 \mathbf{u}
\end{aligned}
\tag{3.131}
$$

where $\boldsymbol{\varepsilon}$, $\mathbf{u}$, and λ, μ are strain tensor, displacement vector, and Lame's constants, respectively. The governing equation is discretized by applying the interaction models of the MPS method for differential operators. The discretized formulation without considering rotation is expressed as

$$
\rho \frac{d\mathbf{v}_i}{dt} = \lambda \frac{d}{n_0} \sum_j \frac{\mathrm{tr}(\boldsymbol{\varepsilon}_j) + \mathrm{tr}(\boldsymbol{\varepsilon}_i)}{|\mathbf{r}_{ij}^0|^2} \mathbf{r}_{ij}^0 w_{ij}^0 + 2\mu \frac{2d}{n_0} \sum_j \frac{\mathbf{u}_j - \mathbf{u}_i}{|\mathbf{r}_{ij}^0|^2} w_{ij}^0,
\tag{3.132}
$$

where d, n_0, $\mathbf{r}_{ij}^0$, and w_{ij}^0 are spatial dimension, the base particle number density, relative position in the material coordinate, and weight function, respectively. The second term on the right hand side is analogical to the spring connection. For the numerical stability the spring connections between particles are helpful because they suppress unrealistic local particle oscillation (Vignjevic et al., 2000). The model was improved by Kondo et al. (2006) to overcome the instability in three-dimensional calculation. Since the spring connection is a straightforward idea, and the particle motions under it is easy to be understood, the method proposed by Song et al. (2005) has been widely applied to industrial problems.

Inagaki et al. (2008a, b) applied the method in nuclear engineering field, and response analysis of a concrete cask was conducted. Chhatkuli et al. (2009) used the method for dynamic tracking of lung deformation so as to reduce the extra dose in radiation therapy. Similar studies were carried out by Ito et al. (2010a, b, 2011), where the images equivalent to 4DCT (four-dimensional computed tomography) were created using the simulation. Shino et al. (2013) simulated the deformation of the chest, and Ookura et al. (2013) simulated the diaphragm respiratory motion. Yoshida (2011) extended the method for failure analysis to predict seismic behavior of ground structures. Fluid—elastic solid interaction problems were solved by Hwang et al. (2014) using their improved MPS method. Slamming was analyzed by Khayyer and Gotoh (2016) and Khayyer et al. (2017) using their improved MPS method.

However, the spring-based model proposed by Song et al. (2005) had a difficulty in extending it to nonlinear structural material. It is because the structural equations are usually expressed by the stress tensor and the strain tensor, which do not appear explicitly in the spring-based model. Therefore extension to wider range of structural materials, the method which calculates the stress tensor and strain tensor explicitly, is favorable. Such a method was proposed by Suzuki and Koshizuka (2007, 2008) for hyperelastic solid analysis using the least square approximation. Suzuki and Koshizuka (2008) adopted the RATTLE algorithm to take the incompressibility condition. It was improved by Kondo et al. (2007, 2010a) where the complementary force to suppress the local unrealistic particle oscillation (Vignjevic et al., 2000) was proposed.

Here, another method for elastic solid (Kondo et al., 2010a) with the explicit algorithm using the least square approximation is going to be explained. The governing equation is Saint Venant-Kirchhoff model:

$$S = 2\mu E + \lambda \mathrm{tr}(E)I, \tag{3.133}$$

where S is the second Piola-Kirchhoff stress, and E is the Green-Lagrange strain defined as

$$E = \frac{1}{2}(F^T F - I). \tag{3.134}$$

F is a deformation gradient tensor expressed as

$$F = \frac{\partial \mathbf{u}}{\partial \mathbf{x}^0}, \tag{3.135}$$

where $\mathbf{u}$ is the displacement from the initial position, and $\mathbf{x}^0$ is the material coordinate for which the initial position is used in practical calculations. Since adjacency to the neighbor particles is maintained in the elastic solid analysis, the weight function is calculated using the initial position of the particles. In specific the weight function is given as

$$w_{ij}^0 = \begin{cases} 1 - \dfrac{|\mathbf{r}_{ij}^0|^2}{r_e^2} & (|\mathbf{r}_{ij}^0| < r_e) \\ 0 & (|\mathbf{r}_{ij}^0| \geq r_e) \end{cases}, \tag{3.136}$$

where r_e is an effective radius, and $\mathbf{r}_{ij}^0$ is the relative positon vector at the initial state. Upper index "0" means the value at the initial state. The deformation gradient tensor can be given using the least square approximation as

$$F_i = \left[\sum_{j \neq i} \mathbf{r}_{ij} \otimes \mathbf{r}_{ij}^0 w_{ij}^0 \right] A_i^{-1}, \tag{3.137}$$

where

$$A_i = \left[\sum_{j \neq i} \mathbf{r}_{ij}^0 \otimes \mathbf{r}_{ij}^0 w_{ij}^0 \right]. \tag{3.138}$$

With this approximation the stress S can be calculated. In addition, the motion equation of the particle is derived from the elastic strain energy:

$$V = \frac{m}{\rho} \sum_i \frac{1}{2} E_i : S_i, \tag{3.139}$$

where m and ρ are the mass of one particle and the density, respectively. The motion equation of the particle is obtained by differentiating the energy V with respect to particle position x. The resulting equation will be

$$\rho \frac{d\mathbf{v}_i}{dt} = \sum_{j \neq i} (F_i S_i A_i^{-1} \mathbf{r}_{ij}^0 + F_j S_j A_j^{-1} \mathbf{r}_{ij}^0) w_{ij}^0. \tag{3.140}$$

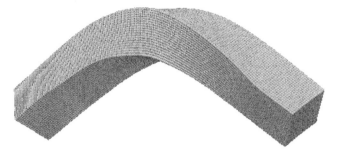

Figure 3.16 Large oscillation of elastic beam (Kondo et al., 2010a).

For time integration a symplectic scheme can easily be adopted as

$$\begin{aligned}
\mathbf{x}_i^{k+1} &= \mathbf{x}_i^k + \mathbf{v}_i^k \Delta t \\
\mathbf{v}_i^{k+1} &= \mathbf{v}_i^k + (d\mathbf{v}/dt)_i^{k+1} \Delta t
\end{aligned} \tag{3.141}$$

Fig. 3.16 shows an example where the large oscillation of elastic beam is calculated (Kondo et al., 2010a). Since this method has a good energy conservation property, the large deformation can be calculated stably.

Kondo et al. (2010b) calculated elastic behavior of the beam, which is made of random packed powder, and further fundamental study was conducted by Mizutani et al. (2011). Shao et al. (2013) applied the method to the fluid–structure interaction problem, where the explicit MPS method was used for the fluid calculation. In medical field Kikuchi et al. (2014, 2015) adopted the hyperelastic model for the swallowing simulation.

In addition, to broaden the application range of the particle method, thin and thick structure models have also been developed. Kondo and Koshizuka (2010) proposed a thin elastic plate model by extending the least square approximation. The model had good energy conservation property because it was built on the Hamiltonian framework and symplectic scheme was adopted. Another thin structure model was proposed by Shao et al. (2012), where the Reissner-Mindlin shell was modeled.

Fluid–thin structure interaction has been analyzed by the MPS method: sloshing in a tank (Chikazawa et al., 1999, 2001) and deformation of a red blood cell in a blood vessel (Tanaka and Koshizuka, 2007; Tsubota and Wada, 2010).

Mistume et al. (2014, 2015) analyzed fluid–structure interaction problems by using the MPS method for fluids and the Finite Element Method for structures. As the wall boundary condition of the MPS method, wall particles were used in Mitsume et al. (2014), while a temporary mirror particle technique was developed in Mistume et al. (2015).

DEM (Discrete Element Method) is a particle method to analyze powder flow (Cundall and Strack, 1979). Powder–liquid interaction problems have been solved by coupling DEM and the MPS method (Gotoh et al., 2003; Sakai et al., 2008; Zhang et al., 2010; Yamada and Sakai, 2013; Sun et al., 2014).

REFERENCES

Arai, J., Koshizuka, S., 2009. Numerical analysis of droplet impingement on pipe inner surface using a particle method. J. Power Energy Syst. 3, 228–236.

Carlson, M., Mucha, P.J., Turk, G., 2004. Rigid fluid: animating the interplay between rigid bodies and fluid. ACM Trans. Graphics 23, 377–384.

Chhatkuli, S., Koshizuka, S., Uesaka, M., 2009. Dynamic tracking of lung deformation during breathing by using particle method. Modell. Simul. Eng. 2009, Article ID 190307.

Chikazawa, Y., Koshizuka, S., Oka, Y., 1999. Numerical analysis of sloshing with large deformation of elastic walls and free surfaces using MPS method. Trans. Jpn. Soc. Mech. Eng. (B) 65, 2954–2960 [in Japanese].

Chikazawa, Y., Koshizuka, S., Oka, Y., 2001. Numerical analysis of three-dimensional sloshing in an elastic cylindrical tank using Moving Particle Semi-implicit method. Comput. Fluid Dyn. J. 9, 376–383.

Cundall, P.A., Strack, O.D.L., 1979. A discrete numerical model for granular assemblies. Géotechnique 29, 47–65.

Gotoh, H., Hayashi, M., Ando, S., Sakai, T., 2003. Lagrangian coupling for solid-liquid Two phase flow by DEM-MPS method. Ann. J. Hydraul. Eng., JSCE 47, 547–552 [in Japanese].

Hairer, E., Lubich, C., Wanner, G., 2000. Geometric Numerical Integration. Springer.

Heo, S., Koshizuka, S., Oka, Y., 2002. Numerical analysis of boiling on high heat-flux and high subcooling condition using MPS-MAFL. Int. J. Heat Mass Transfer 45, 2633–2642.

Hirt, C.W., Nichols, B.D., 1980. Adding limited compressibility to incompressible hydrocodes. J. Comput. Phys. 34, 390–400.

Hu, F., Matsunaga, T., Tamai, T., Koshizuka, S., 2017. An ALE particle method using upwind interpolation. Comput. Fluids 145, 21–36.

Hwang, S.-C., Khayyer, A., Gotoh, H., Park, J.-C., 2014. Development of a fully Lagrangian MPS-based Coupled Method for simulation of fluid-structure interaction problems. J. Fluids Struct. 50, 497–511.

Ikeda, H., Koshizuka, S., Oka, Y., Park, H.S., Sugimoto, J., 2001. Numerical analysis of jet injection behavior for fuel-coolant interaction using Particle method. J. Nucl. Sci. Technol. 38, 174–182.

Inagaki, K., Sakai, M., Koshizuka, S., 2008a. Response analysis of a concrete cask subjected to seismic motions by MPS method. Trans. Jpn. Soc. Comput. Eng. Sci. 2008, Paper No. 20080026. [in Japanese].

Inagaki, K., Sakai, M., Koshizuka, S., 2008b. Development of a Particle method for dynamic elastic-plastic analysis. Trans. Jpn. Soc. Comput. Eng. Sci. 2008, Paper No. 20080031. [in Japanese].

Inutsuka, S., 2002. Reformulation of smoothed particle hydrodynamics with Riemann solver. J. Comput. Phys. 179, 238−267.

Ito, H., Koshizuka, S., Nakagawa, K., Haga, A., 2010a. Particle simulation of lung deformation in axial plane by chest respiration and heatbeat. Trans. Jpn. Soc. Simul. Technol. 2, 93−100 [in Japanese].

Ito, H., Koshizuka, S., Shino, R., Haga, A., Yamashita, H., Onoe, T., Nakagawa, K., 2010b. Generation method for simulation-based chest 4DCT based on particle simulation for radiotherapy. Med. Imaging Technol. 28, 229−236 [in Japanese].

Ito, H., Koshizuka, S., Haga, A., Nakagawa, K., 2011. Rib Cage Motion Model construction based on patient-specific CT images between inhalation and exhalation. Med. Imaging Technol. 29, 208−214 [in Japanese].

Khayyer, A., Gotoh, H., 2016. A multiphase compressible-incompressible particle method for water slamming. Int. J. Offshore Polar Eng. 26, 1−6.

Khayyer, A., Gotoh, H., Falahaty, H., Shimizu, Y., Nishijima, Y., 2017. Towards development of a reliable fully-Lagrangian MPS-based FSI solver for simulation of 2D hydrostatic slamming. Ocean Syst. Eng. 7, 299−318.

Kikuchi, T., Michiwaki, Y., Koshizuka, S., Kamiya, T., Osada, T., Jinnno, N., Toyama, Y., 2014. Simulation of uniaxial compression based on Hamiltonian MPS method with wall boundary condition using Penalty method. Trans. Jpn. Soc. Comp. Eng. Sci. 2014, Paper No. 20140010.

Kikuchi, T., Michiwaki, Y., Kamiya, T., Toyama, Y., Tamai, T., Koshizuka, S., 2015. Human swallowing simulation based on videofluorography images using Hamiltonian MPS method. Comput. Part. Mech. 2, 247−260.

Kondo, M., Koshizuka, S., Suzuki, Y., 2006. Application of symplectic scheme to three-dimensional elastic analysis using MPS method. Trans. Jpn. Soc. Mech. Eng. (A) 72, 65−71 [in Japanese].

Kondo, M., Suzuki, Y., Koshizuka, S., 2007. Suppressing local oscillation for elastic analysis based on least square approximation using particles. Trans. Jpn. Soc. Comput. Eng. Sci. 2007, Paper No.20070031. [in Japanese].

Kondo, M., Koshizuka, S., 2010. Development of Thin Plate Model using Hamiltonian Particle method. Trans. Jpn. Soc. Comput. Eng. Sci. 2010, Paper No. 20100016. [in Japanese].

Kondo, M., Suzuki, Y., Koshizuka, S., 2010a. Suppressing local particle oscillations in Hamiltonian Particle method for elasticity. Int. J. Numer. Meth.n Eng. 81, 1514−1528.

Kondo, M., Yamada, Y., Sakai, M., Koshizuka, S., 2010b. Development of a particle method for an elastic body considering random packing of powder. J. Soc. Powder Technol. Jpn. 47, 531−538 [in Japanese].

Koshizuka, S., Oka, Y., 1996. Moving-Particle Semi-implicit method for fragmentation of incompressible fluid. Nucl. Sci. Eng. 123, 421−434.

Koshizuka, S., Nobe, A., Oka, Y., 1998. Numerical analysis of breaking waves using the Moving Particle Semi-implicit method. Int. J. Numer. Meth. Fluids 26, 751−769.

Koshizuka, S., Ikeda, H., Oka, Y., 1999. Numerical analysis of fragmentation mechanisms in vapor explosions. Nucl. Eng. Des. 189, 423−433.

Koshizuka, S., Yoon, H.Y., Yamashita, D., Oka, Y., 2000. Numerical analysis of natural convection in a square cavity using MPS-MAFL. Comput. Fluid Dyn. J 8, 485−494.

Leimkuhler, B., Reich, S., 2004. Simulating Hamiltonian Synamics. Cambridge University Press.

Leonard, B.P., 1979. A stable and accurate convective modelling procedure based on quadratic upstream interpolation. Comput. Meth. Appl. Mech. Eng. 19, 59−98.

Mizutani, S., Sakai, M., Shibata, K., 2011. Fundamental study on a structure simulation for the power products. J. Soc. Powder Technol. Jpn. 48, 464−472 [in Japanese].

Mitsume, N., Yoshimura, S., Murotani, K., Yamada, T., 2014. MPS-FEM partitioned coupling approach for fluid-structure interaction with free surface flow. Int. J. Comput. Meth. 11, 1350101.

Mitsume, N., Yoshimura, S., Murotani, K., Yamada, T., 2015. Explicitly represented Polygon Wall Boundary Model for explicit-MPS method. Comput. Part. Mech. 2, 73–89.

Monaghan, J.J., 1988. An introduction to SPH. Comput. Phys. Commun. 48, 89–96.

Monaghan, J.J., 1994. Simulating free surface flows with SPH. J. Comput. Phys. 110, 399–406.

Monaghan, J.J., Price, D.J., 2001. Variational principles for relativistic smoothed particle hydrodynamics. Mon. Not. Astron. Soc. 328, 381–392.

Murotani, K., Oochi, M., Fujisawa, T., Koshizuka, S., Yoshimura, S., 2012. Distributed memory parallel algorithm for explicit MPS using ParMETIS. Trans. Jpn. Soc. Comp. Eng. Sci. 2012, Paper No. 20120012. [in Japanese].

Murotani, K., Koshizuka, S., Tamai, T., Shibata, K., Mitsume, N., Yoshimura, S., Tanaka, S., Hasegawa, K., Nagai, E., Fijisawa, T., 2014. Development of hierarchical domain decomposition explicit MPS method and application to large-scale tsunami analysis with floating objects. J. Adv. Simul. Sci. Eng. 1, 16–35.

Oochi, M., Koshizuka, S., Sakai, M., 2010. Explicit MPS algorithm for free surface flow analysis. Trans. Jpn. Soc. Comput. Eng. Sci. 2010, Paper No. 20100013. [in Japanese].

Oochi, M., Yamada, Y., Koshizuka, S., Sakai, M., 2011. Validation of pressure calculation in explicit MPS method. Trans. Jpn. Soc. Comput. Eng. Sci. 2011, Paper No. 20110002. [in Japanese].

Ookura, T., Ito, H., Koshizuka, S., Nomoto, A., Haga, A., Nakagawa, K., 2013. Diaphragm respiratory motion model with ribcage movement. Med. Imaging Technol. 31, 189–197 [in Japanese].

Sakai, M., Koshizuka, S., Toyoshima, I., 2008. Numerical simulation of solid-liquid flows involving free surface by DEM-MPS method. J. Soc. Powder Technol. 45, 466–477 [in Japanese].

Shakibaeinia, A., Jin, Y.-C., 2010. A weakly compressible MPS method for modeling of open-boundary free-surface flow. Int. J. Numer. Meth. Fluids 63, 1208–1232.

Shao, Y., Ito, H., Shibata, K., Koshizuka, S., 2012. An Hamiltonian MPS formulation for Reissner-Mindlin Shell. Trans. Jpn. Soc. Comp. Eng. Sci. 2012, Paper No. 20120013. [in Japanese].

Shao, Y., Yamakawa, T., Kikuchi, T., Shibata, K., Koshizuka, S., 2013. A Three-Dimensional Coupling Method for fluid-structure interaction problems by using explicit MPS method and Hamiltonian MPS Method. Trans. Jpn. Soc. Comput. Eng. Sci. 2013, Paper No. 20130004. [in Japanese].

Shibata, K., Koshizuka, S., Sakai, M., Tanizawa, K., 2012. Lagrangian simulations of ship-wave interactions in rough seas. Ocean Eng. 42, 13–25.

Shino, R., Koshizuka, S., Ito, H., Jibiki, T., Liu, L., 2013. Generation of a supine image of the breast using Numerical Simulation of the Particle method from a prone image. Med. Imaging Technol. 31, 240–247 [in Japanese].

Song, M., Koshizuka, S., Oka, Y., 2005. Dynamic analysis of elastic solids by MPS method. Trans. Jpn. Soc. Mech. Eng. (A) 71, 16–22 [in Japanese].

Sun, X.S., Sakai, M., Sakai, M.-T., Yamada, Y., 2014. A Lagrangian-Lagrangian Coupled method for three-dimensional solid-liquid flows involving free surfaces in a rotating cylindrical tank. Chem. Eng. J. 236, 122–141.

Suzuki, Y., Koshizuka, S., 2007. Development of a particle method for nonlinear elasto-dynamics. Trans. Jpn. Soc. Comp. Eng. Sci. 2007, Paper No. 20070001. [in Japanese].

Suzuki, Y., Koshizuka, S., Oka, Y., 2007. Hamiltonian Moving-Particle Semi-Implicit (HMPS) method for incompressible fluid flows. Comput. Meth. Appl. Mech. Eng. 196, 2876–2894.

Suzuki, Y., Koshizuka, S., 2008. A Hamiltonian particle method for non-linear elastody-namics. Int. J. Num. Meth. Eng. 74, 1344−1373.

Szmidt, J.K., 1998. On the Hamilton's principle for perfect fluids. Bull. Pol. Acad. Sci. Tech. Sci. 46, 199−207.

Tamai, T., Shibata, K., Koshizuka, S., 2013. Development of the higher-order MPS method using the Taylor Expansion. Trans. Jpn. Soc. Comput. Eng. Sci. 2013, Paper No. 20130003. [in Japanese].

Tamai, T., Koshizuka, S., 2014. Least squares Moving Particle Semi-implicit method. Comput. Part. Mech. 1, 277−305.

Tanaka, M., Koshizuka, S., 2007. Simulation of red blood cell deformation using a Particle Method. Nagare 26, 49−55 [in Japanese].

Tsubota, K., Wada, S., 2010. Effect of the natural state of an elastic cellular membrane on tank-treading and tumbling of a single red blood cell. Phys. Rev. E 81, 011910.

Tsukamoto, M.M., Cheng, L.-L., Nishimoto, K., 2011. Analytical and numerical study of the effects of an elastically-linked body on sloshing. Comput. Fluids 49, 1−21.

Vignjevic, R., Campbell, J., Libersky, L., 2000. A treatment of zero-energy modes in the Smoothed Particle Hydrodynamics Method. Comput. Meth. Appl. Mech. Eng. 184, 67−85.

Xiong, J., Koshizuka, S., Sakai, M., 2010. Numerical analysis of droplet impingement using the Moving Particle Semi-implicit method. J. Nucl. Sci. Technol. 47, 314−321.

Xiong, J., Koshizuka, S., Sakai, M., 2011. Investigation of droplet impingement onto wet walls based on simulation using Particle Method. J. Nucl. Sci. Technol. 48, 145−153.

Xiong, J., Koshizuka, S., Sakai, M., Ohshima, H., 2012. Investigation on droplet impingement erosion during steam generator tube failure accident. Nucl. Eng. Des. 249, 132−139.

Xu, R., Stansby, P., Laurence, D., 2009. Accuracy and stability in Incompressible SPH (ISPH) based on the projection method and a new approach. J. Comput. Phys. 228, 6703−6725.

Yabe, T., Wang, P.-Y., 1991. Unified numerical procedure for compressibleand incompressible fluid. J. Phys. Soc. Jpn. 60, 2105−2108.

Yamada, Y., Sakai, M., Mizutani, S., Koshizuka, S., Oochi, M., Murozono, K., 2011. Numerical simulation of three-dimensional free-surface flows with explicit Moving Particle Simulation method. Trans. At. Energy Soc. Jpn. 10, 185−193 [in Japanese].

Yamada, Y., Sakai, N., 2013. Lagrangian-Lagrangian simulation of solid-liquid flows in a bead mill. Powder Technol. 239, 105−114.

Yoon, H.Y., Koshizuka, S., Oka, Y., 1999a. A Particle-Gridless Hybrid method for incompressible flows. Int. J. Numer. Meth. Fluids 30, 407−424.

Yoon, H.Y., Koshizuka, S., Oka, Y., 1999b. A Mesh-Free Numerical method for direct simulation of gas-liquid phase interface. Nucl. Sci. Eng. 133, 192−200.

Yoon, H.Y., Koshizuka, S., Oka, Y., 2001. Direct calculation of bubble growth, departure and rise in nucleate boiling. Int. J. Multiphase Flow 27, 277−298.

Yoshida, I., 2011. Basic study on failure analysis with using MPS method. J. Jpn. Soc. Civil Eng., Ser. A2 (Applied Mechanics (AM)) 67, 93−104 [in Japanese].

Zhang, S., Morita, K., Shirakawa, N., Yamamoto, Y., 2010. Development of a computational framework on fluid-solid mixture flow simulations for the COMPASS code. J. Power Energy Syst. 4, 126−137.

FURTHER READING

Kikuchi, T., Michiwaki, Y., Koshizuka, S., Kamiya, T., Toyama, Y., 2017. Numerical simulation of interaction between organs and food bolus during swallowing and aspiration. Comp. Biol. Med. 80, 114−123.

CHAPTER 4

Boundary Conditions

Abstract

The particle methods, including the Moving Particle Semi-implicit method, can treat boundaries with large deformation, such as those in free-surface flow problems. However, the numerical treatment of boundary condition is generally quite cumbersome due to the particle deficiency problem. Today, numerical algorithms for the boundary treatment have been developed by a number of studies, but are still the subject of active research. This chapter introduces numerical methods for three types of boundaries: (1) solid wall, (2) free surface, and (3) inlet/outlet boundary. For the solid wall boundary, wall particle method, mirror particle method, polygon wall model, and boundary integral-based method are explained in detail. Advantages and disadvantages are also discussed for each method. For the free-surface boundary, several approaches for the free-surface particle detection and pressure calculation with improved stability and accuracy are described. For the inlet/outlet boundary, representation of particle inflow and outflow is provided.

Keywords: Computational fluid dynamics; Lagrangian meshfree method; moving boundary problem; solid wall; free surface; open boundary; particle deficiency; free-surface detection; particle boundary representation; polygon boundary representation

Contents

4.1 Introduction   155
4.2 Solid Wall   159
    4.2.1 Wall Particle Representation   162
    4.2.2 Mirror Particle Representation   168
    4.2.3 Distance Function—Based Polygon Representation   176
    4.2.4 Boundary Integral—Based Polygon Representation   180
4.3 Free Surface   189
    4.3.1 Free-Surface Particle Detection   193
    4.3.2 Pressure Calculation   200
4.4 Inlet and Outlet Boundary Modeling   207
References   212

4.1 INTRODUCTION

Boundary conditions are practically essential for defining a problem and, at the same time, of primary importance in computational fluid dynamics.

Moving Particle Semi-implicit Method
DOI: https://doi.org/10.1016/B978-0-12-812779-7.00004-7

155

It is because the applicability of numerical methods and the resultant quality of computations can critically be decided on how those are numerically treated. Today, demand for the computational analysis of systems with moving boundary has increased and become more complicated. From such backgrounds, the particle simulation technologies, whose one of the significant features is a high adaptability in modeling boundaries with violent motion, have received increasing attention as a new promising computational paradigm in various fields of science and engineering.

Many important problems such as flow driven by moving objects, free-surface flow, flow involving air bubbles, flow accompanying phase transition, and fluid—structure interaction, are moving boundary problems. In dealing with such boundaries with movement or deformation, the traditional mesh methods such as finite difference method, finite element method, and finite volume method generally encounter difficulties in calculating accurately the geometric shape of the boundary.

In moving mesh-based methods, boundaries are represented explicitly by a conformal mesh. It may be an unstructured grid or a structured grid with coordinate transformation (Ryskin and Leal, 1983). Changes in the boundary shape are expressed directly by displacement of grid points. With this framework, accurate surface tracking can be achieved, while one may suffer from mesh-distortion issues. Mesh distortion can be relaxed, for example, by adopting the arbitrary Lagrangian-Eulerian (ALE) formulation (Hirt et al., 1974; Huerta and Liu, 1988). However, in many cases with large deformation, for example, free-surface flow, mesh distortion remains even with the ALE approach, and it can terminate the computation or introduce significant numerical errors. Other Lagrangian approaches for interface tracking include the front-tracking method (Unverdi and Tryggvason, 1992a) and the immersed boundary method (Peskin, 1972). In the approaches, extra surface mesh that moves in a Lagrangian manner is used to represent moving boundary in addition to fixed main computational mesh. Although mesh-distortion problems can be avoided, a sophisticated procedure (Unverdi and Tryggvason, 1992b) is required to improve robustness and flexibility against large deformation, especially in three-dimensional problems.

Various methods have been devised for the interface capturing method, which is an Eulerian boundary-tracking method. Representative examples include the widely used volume of fluid method (Hirt and Nichols, 1981). These methods identify the interface location implicitly based on a scalar function such as a color function (volume fraction) or a distance function, often referred to as an index function. The surface deformation is described by means of the advection equation of the index function. It is noteworthy that computations are stable even in the presence of large deformation owing to the use of stationary mesh. However, the occurrence of numerical diffusion is inevitable for solving the advection equation. As a result, initially defined sharp interface is getting smeared out as time advances, or volume (mass) conservation law is violated. To address this problem, numerous techniques have been developed. Discretization schemes of convection term with higher order of accuracy and with higher resolution are introduced to reduce numerical diffusion. Methods for reconstruction of interface (Rider and Kothe, 1998) or reinitialization of the index function (Sussman, 2003) are developed to recover interface sharpness. In contrast to the Lagrangian interface tracking, the interface capture method requires an advanced algorithm or an increase of calculation cost in order to accurately calculate the boundary shape. Nevertheless, due to its excellent versatility, the interface capture method has been applied to a wide range of moving boundary problems.

As described earlier, a variety of methods have been developed to deal with moving boundaries in the mesh methods. In any case, the methods require highly sophistication of algorithm or considerable computational expense to achieve an accurate simulation of large boundary deformation. On the other hand, the particle method has some remarkable advantages over the traditional mesh methods. In the meshfree approach, as described also in previous chapters, the connectivity between calculation points (particles) is not prescribed by computational mesh but dynamically constructed based upon the particles' spatial distribution in part of the computation. Since simulations can be run without computational mesh, a sophisticated costly preprocess for mesh generation is not required for the particle method. Furthermore, as particles move in Lagrangian manner, deformable interfaces and moving boundaries can naturally be traced, wherein even the large deformation with topological changes can be handled straightforwardly without special treatment.

The particle method can deal with large deformation of the boundary shape. This feature can be particularly suitable for fluid flow problems with deformable boundary such as fluid—fluid interface. In fact, numerous numerical analyses have been performed to study problems with large interfacial deformation such as free-surface flows and multiphase flows. However, besides the aforementioned valuable characteristics, the Lagrangian meshless calculation algorithm of the particle method also has a serious negative aspect: difficulty in the treatment of fixed boundary (e.g., solid wall). In the meshfree framework of calculation, spatial derivatives of the physical

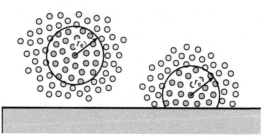

Figure 4.1 Effective domain defined around a fluid particle and particle deficiency encountered near a boundary.

quantities (e.g., velocity and pressure) and value of the particle number density are calculated by referencing the relative positions of surrounding particles that present inside the effective domain, but for particles located near boundary, the effective domain is truncated by the boundary as illustrated in Fig. 4.1. In such situation, for example, the particle number density cannot be properly estimated by the simple summation of weight function over neighboring particles. This is called particle deficiency problem. Therefore, an additional boundary treatment is indispensable not only for imposition of the physical boundary condition, but also for proper management of the near-boundary particle distribution and preventing particles from unphysical behaviors, for example, penetration through impermeable wall. To solve this problem, various ways to represent boundaries have been proposed. Nevertheless, it has still been a challenge to establishing a versatile reliable algorithm offering appropriate treatment of fixed boundary for particle methods.

Boundaries that appear in fluid flow simulations using the Moving Particle Semi-implicit (MPS) method can mostly be classified into following four types: (1) solid wall, (2) free surface, (3) inlet/outlet boundary, and (4) periodic boundary.

The first two of these are frequently used and play major roles in a wide range of problems. The solid wall represents an impermeable surface of solid objects. The geometric shape is commonly presumed to be fixed over time, or rotation, and translation of the rigid body is prescribed as one of the problem configurations. The free surface is a deformable interface between two distinct fluid phases, which are in the usual case liquid and gas. For the gas phase, we often apply an assumption of vacuum, by which the presence of gas is considered negligible. This idealization significantly reduces complexity of the problem.

The inlet boundary is to represent inflow conditions, in which total mass flow rate or velocity profile is often specified. The outlet boundary defines an exit of flow, on which usually pressure is fixed. These boundary conditions, so-called open boundary conditions, are relatively rare in the MPS method as well as other particle methods. That is because the methods suffer from several difficulties in their implementations. For such reasons, the periodic boundary is widely used. Periodic boundary conditions can represent repeated computational domains with ends connected to each other. In combination with a constant driving force, systems with open boundaries, for example, pressure-induced flow in a long duct, can be simulated. However, the periodic boundary also has several drawbacks.

In this chapter, we explain numerical algorithms for the boundary treatment in the MPS method. Thanks to the similarity in the calculation algorithm, many of the techniques proposed for SPH (smoothed particle hydrodynamics) are also applicable in the MPS method. What we introduce here include ones that originally developed to be used in SPH. In Section 4.2, methods for representation of solid wall are presented. In Section 4.3, treatment of free surface is explained, where determination of free-surface particle and techniques for pressure calculations are introduced. In Section 4.4, implementation of inlet and outlet boundaries is described.

4.2 SOLID WALL

Among several kinds of physical boundary conditions in fluid dynamics, solid wall is the most fundamental and important. It represents an impermeable surface of solid objects. Its shape is commonly presumed to be fixed over time (rigid body), while sometimes rotation and translation of the body is prescribed as one of the configurations of problem. In the traditional mesh-/grid-based methods, implementation of solid wall boundary condition is relatively straightforward if the boundary is motionless. On the other hand, in the particle methods including the MPS method, it is often quite cumbersome even for stationary wall. This is due to the particle deficiency problem.

For implementation of solid wall boundaries, we need to adopt an appropriate way to represent boundaries. The boundary representation in the particle method plays two important roles: (1) enforcement of physical boundary condition, (2) compensation of particle deficiency. The former is to impose boundary conditions for physical variables, for example, velocity and pressure. For example, nonslip or free-slip condition for velocity and Neumann boundary condition for pressure are commonly considered on wall. By contrast, the latter involves serious difficulties especially when boundary shape is complicated. These two roles are intimately related to each other, and the roles cannot be considered separately.

In the MPS method, spatial derivatives of the physical quantities and the particle number density are discretized and numerically evaluated by the use of neighboring particles that, presumably, are distributed uniformly. However, for particles located near boundary, the effective domain is truncated by the boundary. This leads to incorrect calculation of particle—wall interaction forces as well as of the particle number density. As the MPS method relies on the particle number density to maintain quasi-homogeneous particle distribution, which is a key to robust particle computation, such an improper estimation can give an immediate deterioration of computation quality. It often appears as unphysical behaviors of near-wall particles. For example, particle can penetrate through wall boundary, or numerical cavity can occur near the surface.

For those reasons, compensation of the particle deficiency has significant importance, while at the same time, it is one of the most difficult parts in the numerical treatment of boundary conditions in the particle methods. Under such circumstances, a number of techniques have been proposed for the boundary representation. Nevertheless, any of them has advantages and disadvantages, and one needs to choose an appropriate representation method depending upon the problem of interest.

This section introduces from commonly used basic to recently developed advanced techniques for treatment of the solid wall. Although there exists a variety of ways for the implementation, in most MPS simulations the solid wall boundaries have been represented using either particles or polygons (see Fig. 4.2). We call the former "particle boundary representation" and the latter "polygon boundary representation." In either case, discretization schemes for the spatial derivatives and the particle number density can be expressed in a general form:

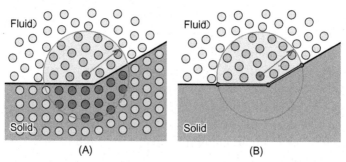

(A)   (B)

Figure 4.2 Boundary representations for solid wall. (A) Particle boundary representation; (B) Polygon boundary representation.

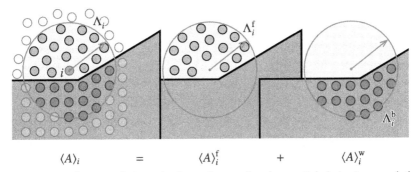

$$\langle A\rangle_i \qquad = \qquad \langle A\rangle_i^{\mathrm{f}} \qquad + \qquad \langle A\rangle_i^{\mathrm{w}}$$

Figure 4.3 Evaluation of discretization schemes for the spatial derivatives and the particle number density near wall.

$$\langle A\rangle_i = \langle A\rangle_i^{\mathrm{f}} + \langle A\rangle_i^{\mathrm{w}} \qquad (4.1)$$

where superscripts f and w indicate the contribution from fluid and wall phases, respectively. The fluid term $\langle A\rangle_i^{\mathrm{f}}$ can be simply evaluated using the MPS discretization schemes (presented in Chapter 2: Fundamental of Fluid Simulation by the MPS Method). The wall term $\langle A\rangle_i^{\mathrm{w}}$ is an additional term for near-wall particles to consider the presence of solid wall, as illustrated in Fig. 4.3. Although they often result in a similar form of discrete equation as the fluid term, the specific formula is decided on how the solid wall is represented. It should be noted that the MPS discretization schemes can be split into two parts in this manner, but this is not always true for other schemes, for example, ones based on the least squares method. However, this does not mean that techniques to be introduced here are unavailable for those exceptions.

The particle boundary representations use additional particles that are usually placed either in the solid domain or on the surface. We call them "boundary particles." Although the detailed ways to define the boundary particle and to discretize the governing equations differ depending on the method, the particle boundary representations can typically express the contribution from wall phase as the summation over neighbor boundary particles:

$$\langle A\rangle_i^{\mathrm{w}} = \sum_{j\in\Lambda_i^{\mathrm{b}}} B_{ij}\ w\big(\big|\boldsymbol{r}_j - \boldsymbol{r}_i\big|\big) \qquad (4.2)$$

where Λ_i^{b} denotes a set of neighbor boundary particles, and w the weight function. B_{ij}, a function of relative position and physical quantity difference between particles i and j, is commonly equivalent to that in the fluid contribution term. Therefore, the particle boundary representations are characterized for its simplicity in discretization once boundary particles are defined. However, determining appropriate distribution of boundary particles for arbitrary shapes is an open question. Furthermore, the inevitable surface roughness may become a source of additional computational error. In this section, we describe two representative methods that categorized into the particle boundary representation: "wall particle representation" and "mirror particle representation."

The polygon boundary representations are relatively new. They use surface mesh (a set of polygons) to represent solid wall. Comparing to boundary particles, surface mesh can be easily handled through use of CAD (computer-aided design) tools without losing sharp details. For their great convenience, the polygon boundary representations have gained considerable attention particularly in engineering fields. However, formulation of the wall contribution terms in discretization schemes is not as straightforward as that in the particle boundary representations. In this section, we introduce two characteristic methods: "distance function−based polygon representation" and "boundary integral−based polygon representation."

4.2.1 Wall Particle Representation

The wall particle method developed by Koshizuka et al. (1998) represents solid bodies by filling their domains with boundary particles that are fixed in the frame of reference of the objects. In order to solve the particle deficiency, thickness of the set of boundary particles needs to be at least as large as the effective radius. Moreover, boundary particles should approximately be distributed uniformly with the same density as in the fluid phase. Those boundary particles that locate within the effective radius from a fluid particle in question are included in the evaluation of spatial derivatives and particle number density. Specifically, the particle number density is calculated by:

$$\langle n \rangle_i = \langle n \rangle_i^{\mathrm{f}} + \langle n \rangle_i^{\mathrm{w}} \tag{4.3}$$

$$\langle n \rangle_i^{\mathrm{f}} = \sum_{j \in \Lambda_i^{\mathrm{f}}} w\big(|r_j - r_i|\big) \tag{4.4}$$

$$\langle n \rangle_i^{\mathrm{w}} = \sum_{j \in \Lambda_i^{\mathrm{b}}} w\big(|r_j - r_i|\big) \tag{4.5}$$

gradient operator:

$$\langle \nabla \phi \rangle_i = \langle \nabla \phi \rangle_i^{\mathrm{f}} + \langle \nabla \phi \rangle_i^{\mathrm{w}} \tag{4.6}$$

$$\langle \nabla \phi \rangle_i^{\mathrm{f}} = \frac{d}{n^0} \sum_{j \in \Lambda_i^{\mathrm{f}}} \frac{\phi_j - \phi_i}{|\boldsymbol{r}_j - \boldsymbol{r}_i|^2} \left(\boldsymbol{r}_j - \boldsymbol{r}_i \right) \; w\!\left(|\boldsymbol{r}_j - \boldsymbol{r}_i| \right) \tag{4.7}$$

$$\langle \nabla \phi \rangle_i^{\mathrm{w}} = \frac{d}{n^0} \sum_{j \in \Lambda_i^{\mathrm{b}}} \frac{\phi_j - \phi_i}{|\boldsymbol{r}_j - \boldsymbol{r}_i|^2} \left(\boldsymbol{r}_j - \boldsymbol{r}_i \right) \; w\!\left(|\boldsymbol{r}_j - \boldsymbol{r}_i| \right) \tag{4.8}$$

Laplacian operator:

$$\langle \nabla^2 \phi \rangle_i = \langle \nabla^2 \phi \rangle_i^{\mathrm{f}} + \langle \nabla^2 \phi \rangle_i^{\mathrm{w}} \tag{4.9}$$

$$\langle \nabla^2 \phi \rangle_i^{\mathrm{f}} = \frac{2d}{\lambda^0 n^0} \sum_{j \in \Lambda_i^{\mathrm{f}}} \left(\phi_j - \phi_i \right) \; w\!\left(|\boldsymbol{r}_j - \boldsymbol{r}_i| \right) \tag{4.10}$$

$$\langle \nabla^2 \phi \rangle_i^{\mathrm{w}} = \frac{2d}{\lambda^0 n^0} \sum_{j \in \Lambda_i^{\mathrm{b}}} \left(\phi_j - \phi_i \right) \; w\!\left(|\boldsymbol{r}_j - \boldsymbol{r}_i| \right) \tag{4.11}$$

Here, Λ_i^{f} indicates a set of neighbor fluid particles around particle i and Λ_i^{b} is a set of neighbor boundary particles ($\Lambda_i = \Lambda_i^{\mathrm{f}} \cup \Lambda_i^{\mathrm{b}}$, $i \notin \Lambda_i$). These equations are written in a general form, and for particles far away from boundary the wall contribution terms $\langle \Box \rangle_i^{\mathrm{w}}$ reduce to zero since Λ_i^{b} becomes empty.

Physical quantities on boundary particles are determined so that pre-scribed boundary conditions are satisfied. Nonslip condition with constant wall velocity $\boldsymbol{u}_{\mathrm{wall}}$ is expressed simply by setting velocity on boundary particles as

$$\boldsymbol{u}_j = \boldsymbol{u}_{\mathrm{wall}} \;\; (j \in \Lambda_i^{\mathrm{b}}) \tag{4.12}$$

Stationary walls can also be expressed in the same manner with $\boldsymbol{u}_{\mathrm{wall}} = \boldsymbol{0}$. On the other hand, for free-slip condition velocity on boundary particles is not uniquely determined. Thus, it is defined virtually only while evaluating discretization schemes. On the free-slip wall, velocity satisfies the condition of wall shear stress

$$(I - \boldsymbol{n} \otimes \boldsymbol{n}) \; \mu \left[(\nabla \boldsymbol{u})^{\mathrm{T}} + \nabla \boldsymbol{u} \right] \; \boldsymbol{n} = 0 \tag{4.13}$$

where I, $\boldsymbol{n}$, and μ are an identity matrix, a unit normal vector on wall, and a viscosity coefficient, respectively. To reproduce such a condition, the viscous term, that is, Laplacian operator for velocity, is evaluated employing

$$\boldsymbol{u}_j = \boldsymbol{u}_i \;\; (j \in \Lambda_i^{\mathrm{b}}) \tag{4.14}$$

which results in

$$\mu\langle\nabla^2\boldsymbol{u}\rangle_i^{\mathrm{w}} = \boldsymbol{0} \tag{4.15}$$

Apparently, Eq. (4.15) represents a condition such that none of viscous forces are applied to fluid particles from wall, which fulfills the free-slip condition of Eq. (4.13). Note that those velocity values defined for the enforcement of boundary conditions are not necessarily linked with the actual movement of boundary particles. For example, the well-known lid-driven cavity flow problem is frequently simulated using spatially fixed boundary particles with finite sliding wall velocity.

By contrast, Neumann boundary condition of pressure requires identification of boundary particles based on their positions. Boundary particles close to body surface (ones exposed to fluid region) are treated as fluid particles while calculating pressure, that is, solving the pressure Poisson equation. We call them "wall particles." The other boundary particles are called "dummy particles" (see Fig. 4.4). Similar to the treatment of the free-slip velocity condition, pressure on dummy particle is virtually defined as

$$P_j = P_i \ (j\in\Lambda_i^{\mathrm{dummy}}) \tag{4.16}$$

where $\Lambda_i^{\mathrm{dummy}}$ is a set of neighbor dummy particles. If we indicate a set of neighbor wall particles by $\Lambda_i^{\mathrm{wall}}$ ($\Lambda_i^{\mathrm{b}} = \Lambda_i^{\mathrm{wall}} \cup \Lambda_i^{\mathrm{dummy}}$), the wall contribution terms in gradient and Laplacian of pressure can be rewritten as

$$\langle\nabla P\rangle_i^{\mathrm{w}} = \frac{d}{n^0}\sum_{j\in\Lambda_i^{\mathrm{wall}}}\frac{P_j - \hat{P}_i}{\left|\boldsymbol{r}_j - \boldsymbol{r}_i\right|^2}(\boldsymbol{r}_j - \boldsymbol{r}_i)\,w\left(\left|\boldsymbol{r}_j - \boldsymbol{r}_i\right|\right) + \frac{d}{n^0}\sum_{j\in\Lambda_i^{\mathrm{dummy}}}\frac{P_i - \hat{P}_i}{\left|\boldsymbol{r}_j - \boldsymbol{r}_i\right|^2}(\boldsymbol{r}_j - \boldsymbol{r}_i)\,w\left(\left|\boldsymbol{r}_j - \boldsymbol{r}_i\right|\right)$$

$$\tag{4.17}$$

$$\langle\nabla^2 P\rangle_i^{\mathrm{w}} = \frac{2d}{\lambda^0 n^0}\sum_{j\in\Lambda_i^{\mathrm{wall}}}(P_j - P_i)\,w\left(\left|\boldsymbol{r}_j - \boldsymbol{r}_i\right|\right) \tag{4.18}$$

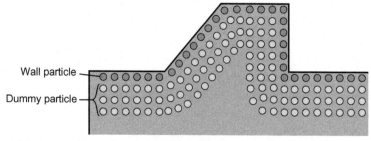

Figure 4.4 Arrangement of wall particles and dummy particles in the wall particle method by Koshizuka et al. (1998).

where the pressure gradient model by Koshizuka and Oka (1996) is applied, and the minimum pressure $\hat{P}_i$ in the gradient model is defined as

$$\hat{P}_i = \min_{j \in \Lambda_i \cup \{i\}} P_j \qquad (4.19)$$

Above-explained wall particle method, developed by Koshizuka et al. (1998), has been used widely and known to be robust. Identification of boundary particles into wall and dummy particles for pressure calculation is known to be effective especially in preventing fluid particles from penetrating wall boundaries. Another strong point is simplicity of the algorithm. Since it can be programmed with relative ease, the method is often implemented in in-house code for the MPS method. However, it has some drawbacks.

Firstly, deployment of boundary particles is considerably cumbersome. For complicated surface shape, equally spaced particle distribution with nearly constant thickness is not readily found. Thus, such preprocess to place boundary particles is operated by the human hand in many cases, and it is usually a time-consuming process, like mesh generation in the grid methods. In addition, since boundary particles should render approximately the same density as in the fluid phase, altering spatial resolution of simulation requires rebuilding them from scratch.

Secondly, there are limitations for the representable solid body shape. As it is represented by a finite number of particles, the sharp details of geometry information are not always reproduced. Even for a simple flat plate, the surface roughness is inevitable and may become a source of additional computational error. Another limitation is for the thickness of rigid body. It has to be larger than the effective radius in order to avoid unphysical interparticle interactions across impermeable boundary. Hence, thin solid plates cannot be considered appropriately without special treatment such as the visibility criterion (Belytschko et al., 1996).

Thirdly, it uses a large amount of memory to store boundary particles. The number of boundary particles increases almost proportionally to the surface area of wall. Thus, three-dimensional problems with complicated geometries use a considerable number of particles just to express solid wall. In such cases, the number of boundary particles is particularly several times larger than that of fluid particles. Consequently, the computation time is also increased more than is necessary.

Lastly, the computational accuracy sometimes becomes an issue. For Dirichlet boundary conditions, like nonslip condition, a physical variable of boundary particles is fixed to a specified value. However, the boundary particles are in general not located on the surface but at a distance equal to or larger than half the particle spacing. Thus, physical quantity at the boundary is not exactly matched to the prescribed value. In case of stationary wall, for example, such a velocity mismatch leads to an artificial slip on wall, which results in an overestimation of flow rate for problems like Hagen—Poiseuille flow. In addition, Neumann boundary conditions also have difficulties. The pressure calculation framework with wall and dummy particles is based on homogeneous Neumann boundary condition

$$\boldsymbol{n} \cdot \nabla P|_{\text{wall}} = 0 \tag{4.20}$$

However, more precisely, the pressure gradient at wall satisfies nonhomogeneous Neumann boundary condition

$$\boldsymbol{n} \cdot \nabla P|_{\text{wall}} = \rho \boldsymbol{f} \cdot \boldsymbol{n} \tag{4.21}$$

where ρ and $\boldsymbol{f}$ denote fluid density and the sum of viscous and external forces, respectively. Note that if the wall moves with finite acceleration, the acceleration term needs to be included

$$\boldsymbol{n} \cdot \nabla P|_{\text{wall}} = \rho \left(\boldsymbol{f} - \frac{\partial \boldsymbol{u}}{\partial t} \bigg|_{\text{wall}} \right) \cdot \boldsymbol{n} \tag{4.22}$$

As pointed out by Hosseini and Feng (2011), artificial homogeneous Neumann condition of Eq. (4.20) produces a numerical boundary layer, and computational accuracy can be diminished. Furthermore, application of Neumann boundary conditions in which constant slope is specified such as heat flux condition is nontrivial.

Some of the issues related to computational accuracy and Neumann boundary conditions, mentioned earlier, can be remedied by utilizing the technique called "fixed ghost particles" proposed by Marrone et al. (2011). Their method uses "interpolation points" located inside fluid domain near surface. The interpolation point itself does not interact physically with fluid particles nor boundary particles, but it has a role to play that improves boundary treatment. An interpolation point is defined in principle per a boundary particle, the location of it is determined to be symmetric to the linked boundary particle with respect to the boundary, as shown in Fig. 4.5. An interpolation point is in fluid domain side, and physical quantities can be interpolated from nearby fluid particles at the point. The interpolant can take different forms; for example, Marrone et al. (2011) adopted the moving least squares approximation (Lancaster and Salkauskas, 1981) to ensure the accuracy, but

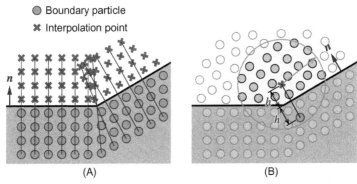

Figure 4.5 Schematic illustration of the fixed ghost particles by Marrone et al. (2011). (A) Interpolation points generated with respect to boundary particles; (B) determination of physical quantities on a boundary particle by the use of interpolation.

low-cost simple weighted average is also available as shown in Asai et al. (2013). The value obtained via the interpolation is used to determine physical quantity on the boundary particle. Specifically, Dirichlet boundary condition can be presented by setting a quantity on boundary particle as

$$\phi_j = 2\phi_{\mathrm{wall}} - \overline{\phi}_j \quad (j \in \Lambda_i^{\mathrm{b}}) \tag{4.23}$$

where $\overline{\phi}_j$ denotes the interpolated quantity at the interpolation point associated to boundary particle j. Similarly, for Neumann boundary condition it is evaluated based on the specified slope and the distance to wall h as

$$\phi_j = \overline{\phi}_j - 2h \, \boldsymbol{n} \cdot \nabla\overline{\phi}\big|_{\mathrm{wall}} \quad (j \in \Lambda_i^{\mathrm{b}}) \tag{4.24}$$

In this way, the mismatch of Dirichlet condition at boundary is resolved, and nonhomogeneous Neumann boundary condition of Eqs. (4.21) and (4.22) can be applied. Furthermore, unlike the classic wall particle method, quantities on each boundary particle are uniquely determined even for Neumann boundary conditions. This is beneficial for simplifying a simulation program. Note that, as long as complexity of boundary shape is moderate enough, generation of interpolation points should not become a significant overhead. This is because it can be performed with a simple calculus once boundary particles are defined. The mirroring process (determination of symmetric position) requires only information of closest point on the body surface. Now let $\boldsymbol{X}$ be the closest position on surface from a boundary particle located at $\boldsymbol{r}_j$. The position of corresponding interpolation point $\boldsymbol{r}_i$ becomes

$$\boldsymbol{r}_i = 2\boldsymbol{X} - \boldsymbol{r}_j \tag{4.25}$$

In addition, the outward unit normal vector n can also be estimated by using X as

$$n = \frac{r_j - X}{|r_j - X|} \qquad (4.26)$$

If, however, the boundary surface has geometric singularities, that is, sharp edges or narrow corners, special treatments must be taken to generate appropriate interpolation points as described in the appendix of Marrone et al. (2011).

Besides these advantages over the classic wall particle method, there are a few disadvantages. Firstly, the computational cost is increased. Not only the position but also the neighbor particle list needs to additionally be stored for every interpolation point. Note that the wall particle method uses the neighbor particle list of wall particles but not of dummy particles. Thus, the computation time as well as the memory usage becomes larger. Secondly, the calculation code becomes more sophisticated. The sophistication is mainly due to the need for geometry calculations. To achieve the automatic determination of interpolation point position from given boundary particle, the closest point search algorithm must be implemented for arbitrary-shaped surface mesh. Lastly, it is difficult to deal with the deformation of solid body. As will be shown later, dynamics of solid materials can also be modeled by the MPS method. Thus, fluid–structure interaction problems can be simulated using the current particle system with additional solving of corresponding governing equations on boundary (solid) particles, where the entire volume of solid phase must be filled with particles. It is rather straightforward for the classic wall particle method to reflect the resulted displacement obtained through the structural analysis by simply migrating particles. However, the fixed ghost particles are required to update their interpolation points as well. This updating is not easily accomplished via the above-explained automated treatment since the deformation must be appropriately expressed also for the surface mesh.

4.2.2 Mirror Particle Representation

The mirror particle method (also referred to as ghost particle method) first introduced by Libersky et al. (1993) is one form of the particle boundary representations. Like the wall particle method, it is based on a strategy using boundary particles with which solid domains are filled, but boundary particles move along with fluid particles near surface. The way to define discretization schemes for the spatial derivatives and the particle number density is basically equivalent to that in the wall particle method; thus, Eqs. (4.3)−(4.11) are available. Main characteristics of this method are related to the way to define the boundary particles, which we call "mirror particles." Each mirror particle represents a mirror image of parent fluid particle. This is similar to the relation between boundary particle and interpolation point seen in the fixed ghost particles (Section 4.2.1). However, a single fluid particle can be a parent of multiple mirror particles, and the number of mirror particles is in general not preliminary given. Moreover, mirroring rules, with which mirror images are defined in accordance, are more complicated. In this section, we first describe the basic rule presented by Cummins and Rudman (1999) and known issues, followed by several techniques remedying them.

The concept is similar to the method of images (Blake, 1971), which is used to satisfy boundary conditions in the analysis of Stokes flows. The mirror particle method represents wall boundary by using mirror particles whose positions are defined by reflecting the fluid particles across the boundary position. Fig. 4.6 shows an example of mirror particles generated near a flat wall. Each fluid particle within the effective radius from boundary is mirrored, and corresponding reflected image is defined on the opposite side as a mirror particle. Here, the mirror particle position is easily determined using a plane reflection matrix $(I - 2n \otimes n)$ as

$$r' = (I - 2n \otimes n) \ (r - p) + p \qquad (4.27)$$

where r is the parent fluid particle position, r' is the generated mirror particle position, and p is an arbitrary point on the boundary. On the other hand, for a fluid particle in the vicinity of a corner, an additional mirror particle is generated at the symmetric position about the corner point, as illustrated in Fig. 4.7. Such corner mirror images exist due to systems with multiple mirrors in which ray can bounce multiple times. In general, different ray path can form different mirror image, see Reshetouski and Ihrke (2013) for more detailed description. This fact makes it highly complicated to determine mirror particles for arbitrary boundary shape. However, by restricting the configuration of boundaries to be a special case in which mirror particles are uniquely defined independent of the origin of the ray, the mirror particle method can be applicable with an affordable computational cost. That is, like an example shown in Fig. 4.7, any interior angle being $\pi/2$. This restriction makes calculation

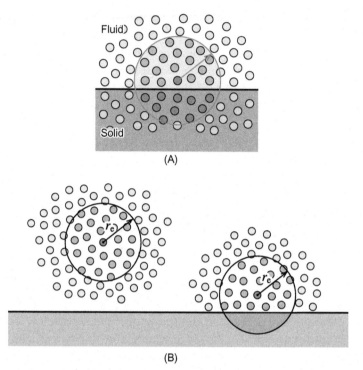

Figure 4.6 Representation of a flat solid wall in the mirror particle method. (A) Mirror particles generated with respect to the wall; (B) velocity vectors assigned to mirror particles for no-slip and free-slip boundary conditions.

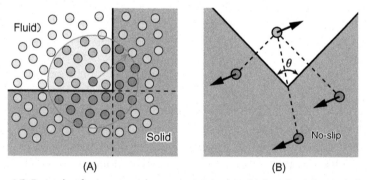

Figure 4.7 Example of mirror particle generation in the vicinity of a right-angled corner based on the rule by Cummins and Rudman (1999). (A) Arrangement of mirror particles; (B) velocity vectors assigned to mirror particles for no-slip boundary condition.

of mirror particles rather simple, but instead imposes a strict limitation on boundary shape.

Physical quantities on mirror particle are determined based on the linear extrapolation from the linked parent fluid particle. Now, the midpoint between fluid and mirror particles is always at the boundary, and thus, the Dirichlet boundary condition can be represented by setting mirror particle so that the arithmetic mean agrees with a value given by the boundary condition. Therefore, for example, mirror particle velocity u' for nonslip boundary conditions is determined simply by

$$u' = 2u_{\text{wall}} - u \qquad (4.28)$$

where u is velocity on the parent fluid particle. For Neumann boundary conditions, quantities are determined based on the slope. For instance, mirror particle pressure P' is calculated like below.

$$P' = P + (r' - r) \cdot \nabla P|_{\text{wall}} \qquad (4.29)$$

Here, P is pressure on the parent fluid particle, and $\nabla P|_{\text{wall}}$ is the prescribed pressure gradient. However, as pointed out by Bierbrauer et al. (2009), the pressure gradient $\nabla P|_{\text{wall}}$ is unknown in general, while only the normal derivative $(n \cdot \nabla P|_{\text{wall}})$ is given. The simplest treatment as employed by Cummins and Rudman (1999) as well as others (Colagrossi and Landrini, 2003; Szewc et al., 2012) is to mimic the homogeneous Neumann boundary condition of Eq. (4.20) by imposing

$$P' = P \qquad (4.30)$$

Contrastingly, others (Akimoto, 2013; Liu et al., 2013) consider the external body force (e.g., gravity) g and employ

$$P' = P + (r' - r) \cdot \rho g \qquad (4.31)$$

Note again that the acceleration term should be considered if any

$$P' = P + (r' - r) \cdot \rho \left(g - \frac{\partial u}{\partial t} \bigg|_{\text{wall}} \right) \qquad (4.32)$$

Considering the external force and the wall acceleration is important to prevent fluid particles from unphysical behaviors in the vicinity of wall. Otherwise, for example, fluid particles may penetrate the object surface.

In contrast to the wall particle method (Section 4.2.1), the mirror particle method has been applied to a smaller number of cases in the MPS method due mainly to the geometric limitation, while it indeed has several advantages.

Firstly, there is no need to place boundary particles in preprocess. Mirror particles are determined based only on information of near-surface fluid particles and boundary geometry. Thus, the particles can be defined without time-consuming manual manipulations. In addition, there is no change in the calculation result due to the difference in wall particle arrangement.

Another advantage is the reduced computational cost. As already mentioned, quantities on mirror particle are determined by means of the simple linear extrapolation from the parent. Therefore, the presence of mirror particle has no influence on the degree of freedom for the pressure Poisson equation, which is typically the dominant cost in MPS simulations. Moreover, information of mirror particles is, in fact, not required to be stored in memory. That is because the wall contribution term (represented by boundary particles) in discretization schemes can be evaluated instantaneously using neighbor fluid particles. Mitsume et al. (2015) has demonstrated that the wall contributions can efficiently be calculated by applying appropriate transformations.

To simplify the determination of mirror images in systems with multiple boundaries, the above-explained standard rule imposes the restriction on geometry: a boundary has to be flat, and a right-angled corner. This means that the computational domain of problem has to be a rectangle in 2D and a cuboid in 3D. This constraint severely limits the scope of application. In such context, various techniques have been developed aiming at extending applicability of the mirror particle method in terms of wall shape.

Treatment of curved surfaces has been reported in several papers (Colagrossi and Landrini, 2003; Børve, 2011; Barker et al., 2014). Determination of mirror particle position is similar to that for flat boundary, as illustrated in Fig. 4.8. For a circular boundary with radius R and center r_c, the mirror particle will be placed at

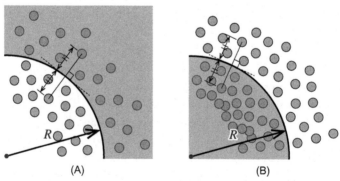

(A)   (B)

Figure 4.8 Mirror particles with respect to a circular boundary with radius R. (A) Concave case ($h > 0$); (B) convex case ($h < 0$).

$$r' = r + \frac{r - r_c}{|r - r_c|} 2h \tag{4.33}$$

where $h = R - |r - r_c|$, which denotes the signed distance of fluid particle to the circle; $h > 0$ denotes a case in which fluid particle locates inside the circle, and $h < 0$ denotes outside. For a more general curved boundary, for example, NURBS (nonuniform rational B-spline) curve or surface, the normal projection of fluid particle onto the boundary will be used to define the reflection point. Like the standard mirroring, the midpoint between fluid and mirror particles is at boundary, and the same procedure (e.g., linear extrapolation) can be taken to determine the physical quantities. However, it should be noticed that through the reflection across a surface with nonzero curvature a volume assigned to each particle should change. The correction factor (the ratio of mirror particle's volume to its parent's volume) can be written in terms of the curvature κ:

$$\beta = \frac{1 + \kappa h}{1 - \kappa h} \tag{4.34}$$

Here, for an arbitrary shape, the curvature should be evaluated at the corresponding reflection point. This equation can be extended to 3D case by using the two principal curvatures κ_1 and κ_2:

$$\beta = \left(\frac{1 + \kappa_1 h}{1 - \kappa_1 h} \right) \left(\frac{1 + \kappa_2 h}{1 - \kappa_2 h} \right) \tag{4.35}$$

Accordingly, the wall contribution term in the discretization schemes should be modified incorporating the correction factor as follows:

$$\langle n \rangle_i^{\mathrm{w}} = \sum_{j \in \Lambda_i^{\mathrm{b}}} \beta_j w \left(|r_j - r_i| \right) \tag{4.36}$$

$$\langle \nabla \phi \rangle_i^{\mathrm{w}} = \frac{d}{n^0} \sum_{j \in \Lambda_i^{\mathrm{b}}} \frac{\phi_j - \phi_i}{|r_j - r_i|^2} (r_j - r_i) \; \beta_j w \left(|r_j - r_i| \right) \tag{4.37}$$

$$\langle \nabla^2 \phi \rangle_i^{\mathrm{w}} = \frac{2d}{\lambda^0 n^0} \sum_{j \in \Lambda_i^{\mathrm{b}}} \left(\phi_j - \phi_i \right) \; \beta_j w \left(|r_j - r_i| \right) \tag{4.38}$$

In this manner, curved surfaces can be treated simply in the mirror particle method. However, this technique has some issues. At first, the way to find a reflection point for an arbitrary shape is nontrivial. As demonstrated by Barker et al. (2014), the closest point search on NURBS curve can be implemented by means of the Newton—Raphson method. However, the implicit calculation entails computational expense, and moreover, the convergence to true solution is not guaranteed due, for example, to the presence of local minimum. Apart from the computational complexity, there is an important limitation on boundary shape. In case of a concave surface, that is, $h > 0$, there can exist more than one closest point when $\kappa \geq 1/h$. Furthermore, in case of a convex surface, that is, $h < 0$, the correction factor will take a negative value when $\kappa > 1/|h|$. It is well known that negative weight function should be avoided since it triggers numerical instability. Accordingly, there is a limitation for the magnitude of curvature:

$$\kappa < \frac{1}{r_e} \tag{4.39}$$

where r_e is the effective radius. Thus, increasing the spatial resolution (reducing the particle spacing) makes it possible to treat boundaries with larger curvature.

Here we discuss about the treatment of a corner with an arbitrary angle. For the sake of simplicity, we here assume that all boundaries are flat. Let θ denote an interior angle between two adjacent boundaries, and take a value within the range of $0 \leq \theta \leq 2\pi$. A corner is concave when $\theta < \pi$, and convex when $\theta > \pi$. The standard mirroring rule, with which all fluid particles in the vicinity of a corner are reflected across the corner point, fails to create a proper mirror particle distribution around a corner, except a special case of $\theta = \pi/2$. To clarify the issues encountered here, some characteristic situations with different angular sizes are depicted in Fig. 4.9. In all cases, a fluid particle is mirrored once for each of three elements: two boundaries and one corner. For a

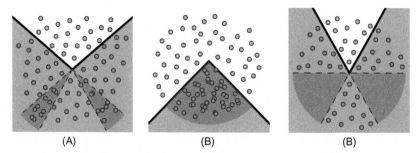

Figure 4.9 Unsuccessful examples of mirror particle generation via the basic mirroring. (A) $\pi/2 < \theta < \pi$; (B) $\theta > \pi$; (C) $\theta < \pi$.

concave corner with $\theta > \pi/2$ as well as a convex corner, a domain into which mirror images generated with respect to a corner overlaps with those for neighbor boundaries. This clearly indicates the existence of nonessential mirror particles, and typically leads to excessive repulsive forces between fluid and wall. On the other hand, with a smaller angle, that is, $\theta < \pi/2$, empty regions appear. This, conversely, suggests a shortage of mirror particles, and may cause penetration of fluid particle through wall.

Various techniques have been invented to deal with corners of arbitrary angle. One possible way to address this problem is to correct weight function of mirror particle by, for example, considering volume change, like the treatment of curved surface. Representative examples are Børve (2011) and Yildiz et al. (2009). Børve (2011) has developed an algorithm to determine the weight function based on the corner angle. Yildiz et al. (2009) have proposed to correct the weight function based on the counted number of times each particle is mirrored. Another strategy is to improve the rule for determination of mirror image itself. Akimoto (2013) has invented a way to eliminate excessive mirror particles for a concave corner with $\theta > \pi/2$ and a convex corner. The algorithm requires solid domain to be decomposed. Then, mirror particle is generated only if the mirrored position is contained inside the domain associated to the boundary in question. Liu et al. (2013) have proposed a different algorithm to remedy excessive mirrors. It does not restrict the generation of mirror particles, but examine the visibility between particles. In the region where mirror particles are overlapped, so-called "over mirror region," a mirror particle is only visible to those present in the same side of domain as its parent. Both approaches are successful to ignore unwanted mirrors, but the method by Liu et al. (2013) outperforms that by Akimoto (2013) in terms of numerical stability. This is because mirror particles shift more continuously in spatial domain in the method by Liu et al. (2013). Matsunaga et al. (2016) have proposed a novel idea for mirror generation. That is to consider mirror particle of mirror particle. By doing so, a shortage of mirror particles encountered for a corner with an angle smaller than $\pi/2$ can be resolved. Matsunaga et al. also generalized the algorithm for the visibility determination so that it can be applied for an arbitrary angle. Owing to the visibility criteria that can eliminate unphysical interparticle connections across impermeable boundary, their method allows thin solid structure to be simulated. While these recent developments have made it possible to deal with corners with an arbitrary angle using the mirror particle method, application to 3D geometry is still a challenge for this method.

4.2.3 Distance Function–Based Polygon Representation

The polygon wall boundary model developed by Harada et al. (2008) is another frequently used technique in the MPS method. In their method, interactions between particle and boundary are estimated from the geometric information, mainly the distance function (distance to the closest point on boundary). Since the surfaces of rigid bodies are represented by surface mesh (a set of polygons), geometry configurations generated using CAD software can straightforwardly be used in simulations. This notable feature highly enhances the convenience especially in dealing with 3D problems containing complicated wall shapes. Today, this technique has received particular interest from engineering fields.

The polygon wall model renders simplified formulae for estimation of the wall contribution terms in discretization schemes. Firstly, wall contribution in the particle number density is approximated by so-called the wall weight function Z

$$\langle n \rangle_i^{\mathrm{w}} = \sum_{j \in \Lambda_i^{\mathrm{b}}} w\big(\big|\mathbf{r}_j - \mathbf{r}_i\big|\big) \approx Z(|\mathbf{r}_{i\mathrm{w}}|) \tag{4.40}$$

The wall weight function is assumed to be a function of the distance to wall $|\mathbf{r}_{i\mathrm{w}}|$, where $\mathbf{r}_{i\mathrm{w}}$ presents the relative position vector pointing the closest point on boundary from given fluid particle i, as illustrated in Fig. 4.10.

The wall weight function is calculated by considering the virtual boundary particles placed outside a flat wall boundary like the conventional wall particle method (see Fig. 4.11). Measured values of the wall

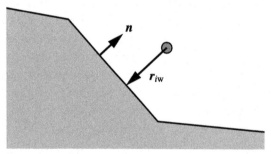

Figure 4.10 Definition of a wall normal vector $\mathbf{n}$, and $\mathbf{r}_{i\mathrm{w}}$ that points the closest point on boundary.

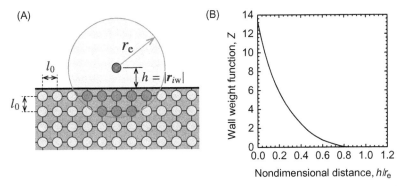

Figure 4.11 Determination of the wall weight function in the polygon wall boundary model by Harada et al. (2008). (A) Arrangement of virtual boundary particles; (B) wall weight function as a function of distance to wall nondimensionalized by effective radius.

weight function by using the virtual boundary particles are stored to construct a look-up table as a function of the distance to wall. During the fluid flow simulation, the wall weight function for a fluid particle in question will be evaluated by interpolating the values stored in the table based on the distance function. Note that although the boundary particles are considered virtually for construction of the look-up table, the wall weight function considered here can avoid issues related to the surface roughness encountered in the wall particle method. This is because a constant arrangement is assumed for the virtual boundary particles, and walls are expressed as smooth surfaces.

The wall contribution in the pressure gradient scheme can approximately be formulated using the weight function as follows:

$$\langle \nabla P \rangle_i^{\mathrm{w}} = \frac{d}{n^0} \sum_{j \in \Lambda_i^{\mathrm{b}}} \frac{P_j - P_i}{\left| \mathbf{r}_j - \mathbf{r}_i \right|^2} \left(\mathbf{r}_j - \mathbf{r}_i \right) \ w \left(\left| \mathbf{r}_j - \mathbf{r}_i \right| \right) \approx \frac{d}{n^0} \frac{P_{\mathrm{wall}} - P_i}{\left| \mathbf{r}_{iw} \right|^2} \mathbf{r}_{iw} \ Z(\left| \mathbf{r}_{iw} \right|)$$

(4.41)

Here, homogeneous constant pressure P_{wall} is assumed on the virtual boundary particles, and applied force is perpendicular to the wall ($\langle \nabla P \rangle_i^{\mathrm{w}}$ is parallel to $\mathbf{r}_{iw}$). Although the distance function ($\left| \mathbf{r}_{iw} \right|$), the wall normal vector ($\mathbf{n} = -\mathbf{r}_{iw}/\left| \mathbf{r}_{iw} \right|$, see Fig. 4.10), and the wall weight function can be obtained by the geometric information, we need to additionally give a condition to determine the wall pressure P_{wall}. The wall normal component of the contact force acting between fluid and wall plays a role to satisfy the impermeability condition. Thus, it can be derived based on the

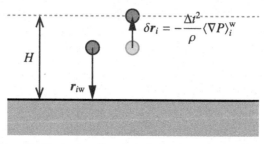

Figure 4.12 Schematic illustration for the relation between wall contact force and position correction.

position-based perspective. A fluid particle that approximated to wall will be pushed away from the wall, like illustrated in Fig. 4.12. This effect is driven by the wall contribution term of the pressure gradient. Therefore, $\langle \nabla P \rangle_i^w$ should be modeled to have enough magnitude to let the particle be back to the certain distance from the wall so that the unphysical particle penetration through solid boundary does not occur. Such repulsive contact forces are considered in the correction step in the MPS method. Recall that the position correction can be described as follows:

$$\delta r_i = -\frac{\Delta t^2}{\rho} \langle \nabla P \rangle_i^w \tag{4.42}$$

Application of the polygon wall approximation of Eq. (4.41) yields

$$\delta r_i = -\frac{\Delta t^2}{\rho} \frac{d}{n^0} \frac{P_{wall} - P_i}{|r_{iw}|^2} r_{iw} \; Z(|r_{iw}|) \tag{4.43}$$

By assuming that the particle will move to the wall normal distance of H after a period of the time step Δt due to the presence of wall, the following relations can be obtained.

$$\delta r_i = \begin{cases} (|r_{iw}| - H)\dfrac{r_{iw}}{|r_{iw}|} & (|r_{iw}| < H) \\[2mm] 0 & (|r_{iw}| \geq H) \end{cases} \tag{4.44}$$

$$\Leftrightarrow P_{wall} = \begin{cases} P_i + \dfrac{\rho}{\Delta t^2} \dfrac{n^0}{d} \dfrac{|r_{iw}|}{Z(|r_{iw}|)}(H - |r_{iw}|) & (|r_{iw}| < H) \\[2mm] P_i & (|r_{iw}| \geq H) \end{cases} \tag{4.45}$$

Substitution of Eq. (4.45) for Eq. (4.43) yields

$$\langle \nabla P \rangle_i^{\mathrm{w}} = \begin{cases} \dfrac{\rho}{\Delta t^2}(H - |\boldsymbol{r}_{iw}|)\dfrac{\boldsymbol{r}_{iw}}{|\boldsymbol{r}_{iw}|} & (|\boldsymbol{r}_{iw}| < H) \\ 0 & (|\boldsymbol{r}_{iw}| \geq H) \end{cases} \qquad (4.46)$$

While value of H is a parameter that can be adjusted depending on problems, $H = l_0/2$ is recommended, where l_0 indicates the particle spacing. This is due to that the virtual boundary particles are arranged as shown in Fig. 4.11, where notice that the first layer of the virtual boundary particle is located not on the surface but at the distance of $l_0/2$.

For the viscosity term, the wall contribution in Laplacian of velocity is approximated again using the wall weight function as

$$\langle \nabla^2 \boldsymbol{u} \rangle_i^{\mathrm{w}} = \frac{2d}{\lambda^0 n^0} \sum_{j \in \Lambda_i^b} (\boldsymbol{u}_j - \boldsymbol{u}_i)\ w\big(|\boldsymbol{r}_j - \boldsymbol{r}_i|\big) \approx \frac{2d}{\lambda^0 n^0}(\boldsymbol{u}_{\mathrm{wall}} - \boldsymbol{u}_i)\ Z(|\boldsymbol{r}_{iw}|)$$

$$(4.47)$$

Here, it is assumed that wall velocity $\boldsymbol{u}_{\mathrm{wall}}$ is applied to all the virtual boundary particles. To enforce the stationary condition, let $\boldsymbol{u}_{\mathrm{wall}}$ be zero.

Similarly, for the pressure Poisson equation, the wall contribution in Laplacian of pressure is approximated as

$$\langle \nabla^2 P \rangle_i^{\mathrm{w}} = \frac{2d}{\lambda^0 n^0} \sum_{j \in \Lambda_i^b} (P_j - P_i)\ w\big(|\boldsymbol{r}_j - \boldsymbol{r}_i|\big) \approx \frac{2d}{\lambda^0 n^0}(P_{\mathrm{wall}} - P_i)\ Z(|\boldsymbol{r}_{iw}|)$$

$$(4.48)$$

where the pressure of the virtual wall particle (P_{wall}) is given by Eq. (4.45). Therefore, this equation can be rewritten as follows:

$$\langle \nabla^2 P \rangle_i^{\mathrm{w}} = \begin{cases} \dfrac{2\rho\,|\boldsymbol{r}_{iw}|}{\lambda^0\,\Delta t^2}(H - |\boldsymbol{r}_{iw}|) & (|\boldsymbol{r}_{iw}| < H) \\ 0 & (|\boldsymbol{r}_{iw}| \geq H) \end{cases} \qquad (4.49)$$

Summarizing the above, the wall contributions in discrete equations (particle number density, pressure gradient term, viscosity term, pressure Poisson equation) are formulated using the wall weight function Z as well as the geometric information such as the distance function and the wall normal vector. The wall weight function can be calculated quite fast by means of the look-up table. However, the other variables, the distance function and the wall normal vector, are also required but with affordable computational cost.

Fluid particles travel as time advances, and the distance to wall also changes from moment to moment. In a case that wall geometry is represented with the limited number of polygons, it may be possible to carry out the closest point search for each fluid particle by the direct approach. However, since calculation of the closest point on a finite-sized polygon entails certain computational expense, the calculation time taken for this procedure becomes unignorable as the number of polygons increases. In this context, the approximated calculation using the background grid has been adopted frequently and known effective. In general, the background grid is constructed preliminary to the simulation, and fixed in the frame of reference of the object. On each grid point, the distance function is stored. Thus, the distance function at an arbitrary position can be evaluated by interpolation. Note that one has no need to generate the background grid that covers whole computational domain but only around the object up to the distance of the effective radius. This is because the wall contributions apparently reduce to zero for $|r_{iw}| \geq r_e$ and the distance function is not required anymore. The wall normal vector can also be calculated using the same background grid, which stores only information of the distance. Here, let the distance function be indicated by $h \ (= |r_{iw}|)$ for the sake of readability. Then, the wall normal vector can be calculated using the gradient of the distance function as

$$\boldsymbol{n} = -\frac{\boldsymbol{r}_{iw}}{|\boldsymbol{r}_{iw}|} = \frac{\nabla h}{|\nabla h|} \tag{4.50}$$

Note that the distance function theoretically fulfills $|\nabla h| = 1$, but numerically $|\nabla h| \neq 1$. This is why the normalization is required. The gradient of distance function can be approximated simply on the grid by means of, for example, the finite difference central differencing scheme and interpolation.

The polygon wall boundary model has been used widely and applied to a large range of the problems. However, unphysical behaviors of near-wall particles are often observed, and ensuring accuracy has been a problem.

4.2.4 Boundary Integral−Based Polygon Representation

The boundary integral−based approaches are polygon boundary representations. Like the distance function−based method, wall geometry is represented by a surface mesh and thus suited for problems with 3D complicated boundaries. The main characteristic is the way to derive the wall contribution term. It is based on the integration over boundary surface. As the boundary integral—based method can consider boundary shape more precisely, resulted computational accuracy is expected to be improved compared to the conventional distance function—based method. For these reasons, these methods have recently attracted attention. However, it should be noted that the boundary integral—based method is relatively new and is still the subject of active research. Thus, it is important to remember that what we introduce here is not something entrenched, but rather under development.

In the SPH method, the boundary integral—based approach has been studied insensibly by several researchers (Feldman and Bonet, 2007; Ferrand et al., 2013; Mayrhofer et al., 2015). However, they apply a variational formulation proposed by Kulasegaram et al. (2004) that gives reformulation of the discretization schemes. It derives equations of motion to evaluate the contact force by reflecting the effects due to the truncation of kernel support by boundary. While their technique is known to be effective to handle arbitrary geometry in SPH, it is not straightforward to apply it in the MPS schemes. Here, giving a focus on the application in the MPS method, we explain the derivation of equations and the calculation techniques used to construct the boundary integral—based method but without the formulation by Kulasegaram et al. (2004).

The boundary integral—based approach was first introduced in the MPS method by Chikazawa et al. (1999). They expressed the wall contribution to the particle number density as

$$\langle n \rangle_i^{\mathrm{w}} = n^0 \frac{\int_{\Omega_i} w(|\boldsymbol{x} - \boldsymbol{r}_i|) \ \mathrm{d}V}{\int_{\Omega^0} w(|\boldsymbol{x} - \boldsymbol{r}_i|) \ \mathrm{d}V} \tag{4.51}$$

where Ω_i denotes the truncated part of the effective domain, and Ω^0 the full effective domain (independent of the presence of wall), as shown in Fig. 4.13. The denominator takes a constant value and can be written, respectively, for 2D and 3D cases as

$$\int_{\Omega^0} w(|\boldsymbol{x} - \boldsymbol{r}_i|) \ \mathrm{d}V = \begin{cases} \int_0^{r_e} 2\pi r w(r) \ \mathrm{d}r & (\text{in 2D}) \\ \int_0^{r_e} 4\pi r^2 w(r) \ \mathrm{d}r & (\text{in 3D}) \end{cases} \tag{4.52}$$

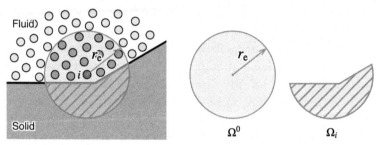

Figure 4.13 Illustration of the full effective domain Ω^0 and the truncated part of the effective domain Ω_i.

Eq. (4.52) can simply be calculated with some calculus. Namely, for the weight function defined by

$$w(r) = \begin{cases} \dfrac{r_e}{r} - 1 & (r < r_e) \\ 0 & (r \geq r_e) \end{cases} \tag{4.53}$$

one finds the following equations

$$\int_{\Omega^0} w(|\boldsymbol{x} - \boldsymbol{r}_i|) \; \mathrm{d}V = \begin{cases} \pi r_e^2 & \text{(in 2D)} \\ \dfrac{2}{3}\pi r_e^3 & \text{(in 3D)} \end{cases} \tag{4.54}$$

Thus, Eq. (4.51) can be rewritten in this form:

$$\langle n \rangle_i^{\mathrm{w}} = \frac{1}{C_V} \int_{\Omega_i} w(|\boldsymbol{x} - \boldsymbol{r}_i|) \; \mathrm{d}V \tag{4.55}$$

where C_V is a constant coefficient that plays a role to properly scale the integral to the desired dimension (C_V itself has the dimension of area in 2D and volume in 3D). Although Eq. (4.51) by Chikazawa et al. (1999) suggests

$$C_V = \begin{cases} \dfrac{\pi r_e^2}{n^0} & \text{(in 2D)} \\ \dfrac{2\pi r_e^3}{3n^0} & \text{(in 3D)} \end{cases} \tag{4.56}$$

we later present a different way to determine C_V to improve the numerical stability.

One has to evaluate the integral in the right-hand side of Eq. (4.55), while the volume integration is not easily handled except for very simple

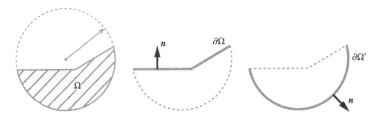

Figure 4.14 Geometric relations of the truncated domain, boundaries, and normal vectors.

shapes, for example, flat boundary. Therefore, the transformation into a form of surface integration is performed. We here present a brief outline for the transformation, for more detailed description refer to the paper by Feldman and Bonet (2007). Let W be a vector function such that

$$\nabla \cdot W = A \qquad (4.57)$$

where A indicates an integrand for volume integral. By the divergence theorem, the following relation holds:

$$\int_\Omega A \ dV = \int_{\partial\Omega} W \cdot n \ dS + \int_{\partial\Omega'} W \cdot n \ dS \qquad (4.58)$$

Here, Ω denotes the truncated domain by wall. $\partial\Omega$ and $\partial\Omega'$ are the boundaries of Ω. n is a unit normal defined on boundaries outward the domain Ω. See Fig. 4.14 for geometric relations. As will be stated later, the vector function W can be selected to satisfy

$$W = 0 \quad \text{on} \quad \partial\Omega' \qquad (4.59)$$

This will simplify the Eq. (4.58) into

$$\int_\Omega A \ dV = \int_{\partial\Omega} W \cdot n \ dS \qquad (4.60)$$

Namely, for the volume integral in Eq. (4.55), one will find a vector function W such that

$$\nabla \cdot W = w(r) \qquad (4.61)$$

If we assume that W can be expressed in a form:

$$W = fx \qquad (4.62)$$

the scalar function f must satisfy the partial differential equation:

$$df + (\boldsymbol{x} \cdot \nabla)f = w \tag{4.63}$$

where d is the number of space dimensions. By further assuming

$$f = f(r) \tag{4.64}$$

where $r = |\boldsymbol{x}|$, Eq. (4.63) can be simplified to the following differential equation:

$$df + r\frac{df}{dr} = w \tag{4.65}$$

$$\Leftrightarrow f = f_d = \frac{1}{r^d} \int r^{d-1} w \ dr \tag{4.66}$$

Therefore, the volume integral can be transformed like the following:

$$\int_\Omega w \ dV = \int_{\partial\Omega} f_d \ \boldsymbol{x} \cdot \boldsymbol{n} \ dS \tag{4.67}$$

which results in the wall contribution of the particle number density in the surface integral form:

$$\langle n \rangle_i^{\mathrm{w}} = \frac{1}{C_V} \int_{\partial\Omega_i} f_d(|\boldsymbol{x} - \boldsymbol{r}_i|) \ (\boldsymbol{x} - \boldsymbol{r}_i) \cdot \boldsymbol{n} \ dS \tag{4.68}$$

Although the specific functional form of f depends on the weight function w, it can be uniquely determined if a boundary condition is given, as will be shown later.

As outlined by Matsunaga et al. (2017), the same procedure can be performed to determine wall contribution for the other schemes, that is, pressure gradient, Laplacian of velocity, and Laplacian of pressure. They can be described in the volume integral form as follows:

$$\langle \nabla P \rangle_i^{\mathrm{w}} = \frac{d}{n^0} \frac{1}{C_V} \int_{\Omega_i} (P - \hat{P}_i) \frac{\boldsymbol{x} - \boldsymbol{r}_i}{|\boldsymbol{x} - \boldsymbol{r}_i|^2} \ w(|\boldsymbol{x} - \boldsymbol{r}_i|) \ dV \tag{4.69}$$

$$\langle \nabla^2 \boldsymbol{u} \rangle_i^{\mathrm{w}} = \frac{2d}{n^0 \lambda^0} \frac{1}{C_V} \int_{\Omega_i} (\boldsymbol{u} - \boldsymbol{u}_i) \ w(|\boldsymbol{x} - \boldsymbol{r}_i|) \ dV \tag{4.70}$$

$$\left\langle \nabla^2 P \right\rangle_i^{\mathrm{w}} = \frac{2d}{n^0 \lambda^0} \frac{1}{C_V} \int_{\Omega_i} (P - P_i) \; w(|\boldsymbol{x} - \boldsymbol{r}_i|) \; \mathrm{d}V \qquad (4.71)$$

Here, the pressure gradient model with minimum pressure is presented to mimic the derivation by Matsunaga et al. (2017), but the application of the boundary integral approach is not limited to this specific scheme. The velocity $\boldsymbol{u}$ and the pressure P at the position $\boldsymbol{x} \in \Omega_i$ for a nonslip surface are given by the boundary conditions as

$$\boldsymbol{u} = \boldsymbol{u}_{\mathrm{wall}} \quad (\boldsymbol{x} \in \Omega_i) \qquad (4.72)$$

$$P = P_i + (\boldsymbol{x} - \boldsymbol{r}_i) \cdot \nabla P \big|_{\mathrm{wall}} \quad (\boldsymbol{x} \in \Omega_i) \qquad (4.73)$$

Note that for clarity, on a free-slip surface $\boldsymbol{u} = \boldsymbol{u}_i$ can be considered, but this apparently results in $\left\langle \nabla^2 \boldsymbol{u} \right\rangle_i^{\mathrm{w}} = 0$. As discussed previously in Section 4.2.2, the pressure gradient on the wall ($\nabla P|_{\mathrm{wall}}$) is indeed unknown in general. Hence, certain presumption should be provided, like for instance Eq. (4.32). Substituting Eq. (4.72) and (4.73) for (4.69) to (4.71) yields

$$\begin{aligned}
\left\langle \nabla P \right\rangle_i^{\mathrm{w}} &= \frac{d}{n^0} \nabla P \big|_{\mathrm{wall}} \cdot \frac{1}{C_V} \int_{\Omega_i} \frac{(\boldsymbol{x} - \boldsymbol{r}_i) \otimes (\boldsymbol{x} - \boldsymbol{r}_i)}{|\boldsymbol{x} - \boldsymbol{r}_i|^2} w(|\boldsymbol{x} - \boldsymbol{r}_i|) \; \mathrm{d}V \\
&+ \frac{d}{n^0} (P_i - \hat{P}_i) \cdot \frac{1}{C_V} \int_{\Omega_i} \frac{\boldsymbol{x} - \boldsymbol{r}_i}{|\boldsymbol{x} - \boldsymbol{r}_i|^2} w(|\boldsymbol{x} - \boldsymbol{r}_i|) \; \mathrm{d}V
\end{aligned} \qquad (4.74)$$

$$\left\langle \nabla^2 \boldsymbol{u} \right\rangle_i^{\mathrm{w}} = \frac{2d}{n^0 \lambda^0} (\boldsymbol{u}_{\mathrm{wall}} - \boldsymbol{u}_i) \cdot \frac{1}{C_V} \int_{\Omega_i} w(|\boldsymbol{x} - \boldsymbol{r}_i|) \; \mathrm{d}V \qquad (4.75)$$

$$\left\langle \nabla^2 P \right\rangle_i^{\mathrm{w}} = \frac{2d}{n^0 \lambda^0} \nabla P \big|_{\mathrm{wall}} \cdot \frac{1}{C_V} \int_{\Omega_i} (\boldsymbol{x} - \boldsymbol{r}_i) \; w(|\boldsymbol{x} - \boldsymbol{r}_i|) \; \mathrm{d}V \qquad (4.76)$$

From the divergence theorem, following general transformations are available:

$$\int_{\Omega} \boldsymbol{x} w \; \mathrm{d}V = \int_{\partial \Omega} \boldsymbol{x} f_{d+1} \; \boldsymbol{x} \cdot \boldsymbol{n} \; \mathrm{d}S \qquad (4.77)$$

$$\int_{\Omega} \frac{x}{|x|^2} w \ dV = \int_{\partial\Omega} \frac{x}{|x|^2} f_{d-1} \ x \cdot n \ dS \tag{4.78}$$

$$\int_{\Omega} \frac{x \otimes x}{|x|^2} w \ dV = \int_{\partial\Omega} \frac{x \otimes x}{|x|^2} f_d \ x \cdot n \ dS \tag{4.79}$$

Thus, we obtain the wall contributions in the surface integral form as follows:

$$\langle \nabla P \rangle_i^w = \frac{d}{n^0} \nabla P|_{\text{wall}} \cdot \frac{1}{C_V} \int_{\partial\Omega_i} \frac{(x-r_i) \otimes (x-r_i)}{|x-r_i|^2} f_d(|x-r_i|) \ (x-r_i) \cdot n \ dS$$
$$+ \frac{d}{n^0}(P_i - \hat{P}_i) \cdot \frac{1}{C_V} \int_{\partial\Omega_i} \frac{x-r_i}{|x-r_i|^2} f_{d-1}(|x-r_i|) \ (x-r_i) \cdot n \ dS$$

$$\tag{4.80}$$

$$\langle \nabla^2 u \rangle_i^w = \frac{2d}{n^0 \lambda^0}(u_{\text{wall}} - u_i) \cdot \frac{1}{C_V} \int_{\partial\Omega_i} f_d(|x-r_i|) \ (x-r_i) \cdot n \ dS \tag{4.81}$$

$$\langle \nabla^2 P \rangle_i^w = \frac{2d}{n^0 \lambda^0} \nabla P|_{\text{wall}} \cdot \frac{1}{C_V} \int_{\partial\Omega_i} (x-r_i) f_{d+1}(|x-r_i|) \ (x-r_i) \cdot n \ dS \tag{4.82}$$

A specific functional for f_m ($m \in \mathbb{Z}$; $\mathbb{Z}$ is the set of integers) can be found by solving

$$f_m = \frac{1}{r^m} \int r^{m-1} w \ dr \tag{4.83}$$

with a boundary condition such that

$$f_m = 0 \quad \text{at} \quad r = r_e \tag{4.84}$$

which greatly simplifies evaluation of the surface integral. This is because, as depicted in Fig. 4.14, only the boundary inside the effective domain will contribute to the calculation and nontruncated part can be neglected. A list of f_m ($m \in [1, 2, 3, 4]$) shows as follows:

$$f_1(r) = \begin{cases} \frac{r_e}{r} \ln\left(\frac{r}{r_e}\right) - 1 + \frac{r_e}{r} & (r < r_e) \\ 0 & (r \geq r_e) \end{cases} \tag{4.85}$$

$$f_2(r) = \begin{cases} \dfrac{r_e}{r} - \dfrac{1}{2} - \dfrac{r_e^2}{2r^2} & (r < r_e) \\ 0 & (r \geq r_e) \end{cases} \qquad (4.86)$$

$$f_3(r) = \begin{cases} \dfrac{r_e}{2r} - \dfrac{1}{3} - \dfrac{r_e^3}{6r^3} & (r < r_e) \\ 0 & (r \geq r_e) \end{cases} \qquad (4.87)$$

$$f_4(r) = \begin{cases} \dfrac{r_e}{3r} - \dfrac{1}{4} - \dfrac{r_e^4}{12r^4} & (r < r_e) \\ 0 & (r \geq r_e) \end{cases} \qquad (4.88)$$

Note again that the scalar functions f_m shown earlier are ones generated for the weight function defined by Eq. (4.53). Thus, if you chose a weight function different than this, you need to derive corresponding f_ms from Eq. (4.83).

The way to determine the scaling factor C_V is nontrivial. Eq. (4.56) can also be a possible choice, but here we present a procedure that gives more numerically conformal settings. That is the use of the wall weight function Z introduced in the distance function—based approach (Section 4.2.3). The wall weight function for a specific distance can be calculated using the virtual boundary particles arranged for a flat boundary. Likewise, the volume integration of the weight function can also be obtained analytically for a flat boundary. Let h be the distance to wall, then the volume integral can be expressed as

$$\int_{\tilde{\Omega}(h)} w \; dV = \begin{cases} 2 \int_h^{r_e} \int_y^{r_e} \dfrac{r}{\sqrt{r^2 - y^2}} w \; drdy & \text{(in 2D)} \\ 2\pi \int_h^{r_e} \int_z^{r_e} rw \; drdz & \text{(in 3D)} \end{cases} \qquad (4.89)$$

where

$$\tilde{\Omega}(h) = \left\{ x \in \mathbb{R}^d \, \middle| \, (0 < |x| < r_e) \wedge (-n \cdot x > h) \right\} \qquad (4.90)$$

If we assume that the weight function is given by Eq. (4.53), they can be determined as

$$\int_{\tilde{\Omega}(h)} w\,dV = \begin{cases} h\sqrt{r_e^2-h^2} + 2r_e h \ln\left(\dfrac{h}{r_e+\sqrt{r_e^2-h^2}}\right) + r_e^2 \arctan\left(\dfrac{\sqrt{r_e^2-h^2}}{h}\right) & \text{(in 2D)} \\[4mm] \dfrac{1}{3}\pi r_e^3\left(1-\dfrac{h}{r_e}\right)^3 & \text{(in 3D)} \end{cases}$$

$$(4.91)$$

The scaling factor C_V is determined so that $\langle n \rangle_i^w$ agrees with the wall weight function at the distance h. This can be satisfied if

$$C_V = \frac{1}{Z(h)}\int_{\tilde{\Omega}(h)} w\,dV \qquad (4.92)$$

From authors' experience, $h = l_0/2$ is one feasible choice, where l_0 indicates the particle spacing.

Through the earlier formulation, each wall contribution term was expressed in the form of the surface integral. In actual simulation, in order to calculate each integral, one has to evaluate the surface integral by part for each polygon element existing in the vicinity of a fluid particle in question. This can be done either numerically or analytically. Various methods are conceivable for the numerical integration procedure. For example, a piecewise quadratic method based on a simple trapezoidal formula can be used. Further study is required to find an efficient integration scheme.

Fig. 4.15 shows simulation example of 2D hydrostatic pressure problem in round-bottom vessel. Here, comparison is done for different representations of solid wall, namely the wall particle method (Koshizuka et al., 1998), the polygon wall boundary model (Harada et al., 2008), and the boundary integral–based method (Matsunaga et al., 2017). It should be found that the polygon wall model shows unnatural particle clustering near the wall surface. On the other hand, the boundary integral–based method agrees well with the wall particle method. Another example is shown in Fig. 4.16, where well-known 2D dam break problem is tested. Again, the polygon wall model presents unphysical behaviors, while the boundary integral–based method indicates good agreement with the wall particle method.

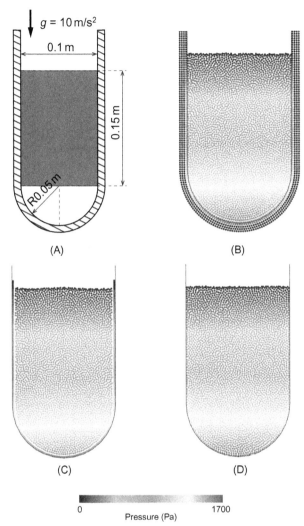

Figure 4.15 Two-dimensional hydrostatic pressure problem in a round-bottom vessel (Matsunaga et al., 2017). (A) Simulation condition; (B) wall particle method; (C) polygon wall model; (D) boundary integral—based method.

4.3 FREE SURFACE

In fluid dynamics, free surface means a deformable interface between two distinct fluid phases. Direct approach for simulating fluid flows with free surface is to treat them as multiphase flow. Numerical simulation of the

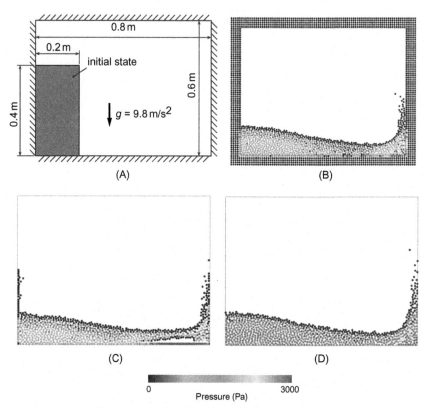

Figure 4.16 Two-dimensional dam break problem (Matsunaga et al., 2017). (A) Simulation condition; (B) wall particle method; (C) polygon wall model; (D) boundary integral−based method.

multiphase flow problem requires equations of motion to be solved for both phases and the solutions to be coupled at the interface to satisfy the boundary condition. Surface tension forces should also be considered at the interface. Such computation can be performed with either of two classes: partitioned or monolithic. Partitioned approaches solve each phase separately, and monolithic approaches solve both phases at one time. However, both approaches have difficulties in calculating flows with large density ratio.

In the MPS method, problems containing multiple fluid materials can be easily handled with a monolithic approach. Different kinds of fluids are treated just by defining them as different kinds of fluid particles.

Interface position is identified implicitly based on changes in the particle type. Governing equations are discretized simply in a coherent way by applying a one-fluid model. Surface tension is approximated as an additional body force, by utilizing the CSF (continuum surface force) models or the potential models as will be presented in Chapter 5, Surface Tension Models in Particle Methods.

However, it is crucial for numerical simulation of multiphase flow to treat carefully the dynamics around the interface. This is due not only to the phenomenological importance or the computational accuracy, but also to the numerical reasons. They are related to the differences in physical properties between the phases. In other words, the gap in the characteristic scales makes it difficult to treat multiple phases together. Typically in problems with large density ratio (say greater than 10), it is common that severe numerical instabilities are observed at the interface. In addition, the pressure Poisson equation becomes ill-conditioned. An iterative linear solver may require a large number of iterations until convergence, or fail to find true solution.

The partitioned approach proposed by Ikeda et al. (1998) addresses this problem by a two-step algorithm for pressure calculation. In the first step, pressure only in a domain of heavier fluid is solved. In doing this, Dirichlet boundary condition is considered at the interface. To this end, neighbor lighter particles are treated like as boundary particles that have constant pressure. In the second step, pressure in a domain of lighter fluid is solved. At this time, Neumann boundary condition is considered at the interface. The partitioned approach can improve the numerical stability, while the computation becomes complicated. An inner loop in the time step loop is necessary to achieve the strong coupling at the interface.

Contrastingly, monolithic approaches are favorable in terms of simplicity of the computational algorithm. In fact, most of the numerical simulations for multiphase flow has been conducted by incorporating the one-fluid model. In order to prevent the numerical instability at the interface, stabilization techniques such as the density smoothing are adopted. Duan et al. (2017a) has succeeded in simulating stably multiphase flows with density ratio of up to approximately 1000 using the MPS method.

Indeed, there are many important problems where multiple phases with large density variations, like liquid and gas, coexist in a system. Typically, liquid is about a thousand times heavier than gas. While in many cases, liquid−gas interactions play an important role in a physical phenomenon of interest, in other cases the presence of gas phase is approximately negligible. In such free-surface flow problems, numerical simulations are conducted without gas particles. It is like assuming the limiting case of infinite density ratio, or in other words, representing circumstances where liquid volume presents in a vacuum. In the absence of gas particles, interactions across the interface cannot be considered. This means, for instance, that effects of entrapped gas on the liquid motion cannot be captured. Therefore, application of this approximation must be carefully examined.

The assumption of vacuum considerably simplifies the computational complexity with respect to the direct simulation of multiphase flow. Since the governing equations are solved only in a liquid phase, occurrence of the numerical instability can be avoided. Furthermore, the calculation cost can be reduced. This is due not only to the reduced number of fluid particles (degree of freedom), but also to the improved conditioning of linear system for the pressure Poisson equation. However, it is not straightforward how to implement free surfaces.

This section puts a focus on the treatment of free surface. While numerical simulation of single-phase free-surface flow might not be as troublesome as that of multiphase flow, implementing free surface is still nontrivial. Treatment of free surface can be divided into two large portions: detection of free-surface particle and imposition of pressure boundary condition.

As no particle presents in the gas phase, fluid particles in the vicinity of free surface can be found based on the reduced particle number density. Such particles are marked as "free-surface particles," as explained previously in Chapter 2: Fundamental of Fluid Simulation by the MPS Method. However, it is known that erroneous determination of free-surface particle occurs frequently with this basic algorithm. Since the inevitable irregular variation of particle distribution can cause an accidental reduction of the particle number density, even particles contained well inside liquid volume can sometimes be misrecognized to be on free surface. Such false free-surface particles are known to produce severe pressure oscillations. Under these circumstances, new algorithms have been developed aiming at improving the identification accuracy. In this section, we present advanced methods that can accurately and efficiently detect free-surface particles.

Boundary condition must be imposed on the free-surface particles. Dirichlet boundary condition is applied for pressure on free surface. Pressure value is, in the usual case, fixed to be zero, which means the atmospheric pressure is assigned on the free surface. Thus, enforcement of the pressure boundary condition on free-surface particles itself is quite straightforward, while in fact, additional numerical treatment is almost necessary. That is to stabilize the computation. As pressure is responsible for recovering the incompressibility, inappropriate pressure distribution can immediately result in unnatural behaviors of particles. In the latter part of this section, we describe several techniques to improve robustness and accuracy in the pressure calculation.

4.3.1 Free-Surface Particle Detection

In the MPS method, the particle number density is used also for the determination of free-surface position. Thanks to the particle deficiency encountered in the vicinity of the free surface, approximation to the boundary is reflected in the measured particle number density. Koshizuka and Oka (1996) proposed a criterion for the particle number density. A particle satisfying the following inequality will be judged to be a free-surface particle.

$$\langle n \rangle_i < \beta n^0 \tag{4.93}$$

Here, β is a threshold for the particle number density. A value of β is usually in the range from 0.9 to 1. With smaller β, the number of free-surface particles to be detected becomes less. Koshizuka (2005) remarked that around $\beta = 0.95$ or 0.97 is recommended based on his experience.

In their model, the calculation algorithm is fairly simple, and moreover, favorable in terms of ease of coding and computation cost without any doubt. However, there is a known issue that false detection of free-surface particles frequently takes place. Although the value of β is a tuning parameter, optimization of the threshold itself cannot be expected to fully resolve the problem. To clarify this, we conducted simple test calculations for examining the identification accuracy with varying β as shown in Fig. 4.17. Here, flow over a step is simulated using the same MPS method but with different free-surface detection algorithm. As shown in Fig. 4.17A for $\beta = 0.97$, obvious false free-surface particles are observed. Conversely, with $\beta = 0.8$ (Fig. 4.17C) the number of free-surface particles is apparently insufficient. Even with a value between these, we could not obtain a calculation result where no misidentification is confirmed, see Fig. 4.17B for the case $\beta = 0.9$.

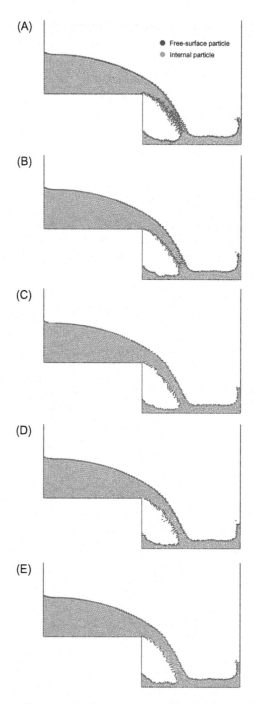

Figure 4.17 Free-surface particle distributions in the simulation of a flow over a step with different detection algorithms. (A) Eq. (4.93) with $\beta = 0.97$; (B) Eq. (4.93) with $\beta = 0.9$; (C) Eq. (4.93) with $\beta = 0.8$; (D) Eqs. (4.93) and (4.94) with $\beta = 0.9$ and $\beta' = 0.8$; (E) Eqs. (4.93) and (4.94) with $\beta = \beta' = 0.9$, and the scanning scheme by Marrone et al. (2010) with $\beta_1 = 0.2$ and $\beta_2 = 0.8$.

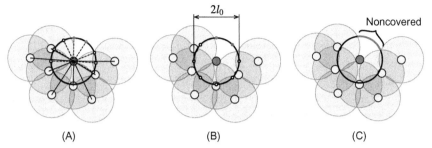

Figure 4.18 Scanning schemes of Dilts (2000). (A) Backpoint scheme; (B) spoke scheme; (C) arc scheme.

Tanaka and Masunaga (2010) proposed a similar approach. In their method, instead of the particle number density, the number of neighbor particles is used for the determination. The proposed criterion can be expressed as

$$N_i < \beta' N^0 \tag{4.94}$$

where N indicates the number of neighbor particles, N^0 the reference value of N, and β' the threshold. Calculation algorithm for this criterion is still quite simple. However, this method can surprisingly improve the determination accuracy if used as an additional condition to Eq. (4.93). Calculation example is shown in Fig. 4.17D, where $\beta = 0.9$ and $\beta' = 0.8$ were used.

Another characteristic approach was proposed by Dilts (2000), where free-surface particle is identified based on a purely geometric condition. The basic idea is to check if a particle is fully covered by its neighboring particles. He devised three scanning schemes to numerically examine whether such a condition is satisfied. Fig. 4.18 illustrates the three schemes: backpoint, spoke, and arc methods. For all of these, a circle of radius being l_0 is defined around each particle. A circle of a particle in question is scanned to examine the coverage. A point on a circle is considered as covered if it is contained by any of neighboring circles. Difference of these three is the way to scan the circle. The backpoint method (Fig. 4.18A) places sampling points at opposite position with respect to each neighbor particle's center position. The spoke method (Fig. 4.18B) distributes them uniformly on the circle. For each sampling point placed, containment must be calculated. In fact, this can be simply accomplished because the containment can be judged based only on the distance to the neighbor particles. On the other hand, the arc method

(Fig. 4.18C) does not calculate the pointwise containment, but instead measures angular intervals of the circular segment overlapped with an adjacent circle. Therefore, the arc method is more complete to find non-covered region than the others. In all schemes, if any noncovered portion is detected, this particle is judged to be a free-surface particle.

The free-surface particle detection based on the scanning scheme by Dilts (2000) is known to perform well. Compared to conventional approaches based on the particle number density or the number of neighbor particles, the determination accuracy will be improved. However, there are severe disadvantages. At first, it requires the computational cost. The geometric operations for checking the containment entail certain calculation expense. In addition, calculation algorithm is complicated compared to the conventional approaches. The algorithmic sophistication is particularly significant when extension of the arc method to 3D space is considered (Haque and Dilts, 2007), where increase in the computational cost is also inevitable.

On the other hand, Marrone et al. (2010) has proposed a new scanning scheme. Like the Dilts' method, the approach finds a free-surface particle based on a geometric condition but with reduced complexity. They utilized the estimation of normal vector to the free surface, and limited the scanning range to this direction. The normal vector can be calculated by equations below:

$$n_i = \frac{v_i}{|v_i|} \tag{4.95}$$

$$v_i = -M_i^{-1} b_i \tag{4.96}$$

$$M_i = \frac{d}{n^0} \sum_{j \in \Lambda_i} \frac{(r_j - r_i) \otimes (r_j - r_i)}{|r_j - r_i|^2} \, w(|r_j - r_i|) \tag{4.97}$$

$$b_i = \frac{1}{n^0} \sum_{j \in \Lambda_i} \frac{r_j - r_i}{|r_j - r_i|} \, w(|r_j - r_i|) \tag{4.98}$$

where d indicates the number of space dimensions, and M is called the moment matrix. To find the normal vector, the inverse of M must be calculated. Once the normal vector is determined, the scan region will be defined as sketched in Fig. 4.19. The criterion for the free-surface particle is that no neighbor particle lies in this parachute-shaped region.

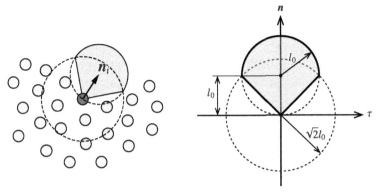

Figure 4.19 Parachute-shaped scan region for scheme of Marrone et al. (2010).

The containment of neighbor particle j in the scan region can be checked via an algorithm shown below:

```
SCAN(r_i,r_j,n_i)
1:      if |r_j − r_i| < √2 l_0 then
2:          if  (r_j − r_i)/|r_j − r_i| · n_i > 1/√2 then
3:              Return TRUE // particle j is in the scan region
4:          end if
5:      else
6:          if |r_j − (r_i + l_0 n_i| < l_0 then
7:              Return TRUE // particle j is in the scan region
8:          end if
9:      end if
10:     Return FALSE // particle j is NOT in the scan region
```

This algorithm is applied to all neighbor particles $j \in \Lambda_i$, and the particle i will be labeled as a free-surface particle if no particle is found in the scan region. Here, it is noteworthy that, despite its geometric standpoint, the criterion proposed by Marrone et al. (2010) is fully computable only with some basic algebraic operations, like inner product and the Euclidian norm. Furthermore, the same algorithm is available even in the 3D case without special modification.

The scanning method by Marrone et al. (2010) outperforms the Dilts' method in terms of computational cost and extension to three dimensions. However, it is still much expensive than the conventional approaches. To remedy this drawback, they suggested to conduct a-priori screening in order to narrow down the candidates of free-surface particles. For the classification of particles, the minimum eigenvalue λ of the moment matrix M is used. The minimum eigenvalue λ_i takes a value in the range of $[0, \langle n \rangle_i / n^0]$ depending upon the spatial organization of neighboring particles. In the vicinity of boundary, the presence of particle deficiency makes λ tend to indicate a smaller value; conversely, an internal particle indicates $\lambda \approx 1$. Therefore, particles can be effectively classified into three groups as follows:

1. $\lambda \leq \beta_1$: free-surface particle,
2. $\beta_1 < \lambda \leq \beta_2$: free-surface candidate, and
3. $\beta_2 < \lambda$: internal particle.

$\beta_1 = 0.2$ and $\beta_2 = 0.75$ are used in Marrone et al. (2010). Only particles classified into group (2) will be reexamined by means of the scanning method. This screening based on the minimum eigenvalue not only reduces the total calculation cost, but also can prevent failure in determination of the scan region. Recall that the inverse of the moment matrix is required to find the normal vector. While the determination of the inverse itself can be completed with the finite number of arithmetic operations, the moment matrix is not always invertible. Degeneration of the moment matrix can happen when the number of neighbor particles is few, like at the tip of fluid volume or free-falling isolated particle.

As performed by Tamai and Koshizuka (2014) and Duan et al. (2017b), further reduction of the computational cost can be achieved, with the same accuracy as Marrone's, by incorporating inexpensive conventional criteria. The idea is to first conduct a rough estimation to narrow down the candidates, and later decide the free-surface particles accurately. For the rough determination, the low-cost simple criterion for the particle number density of Eq. (4.93) by Koshizuka and Oka (1996) is suitable. A criterion for the number of neighbor particles of Eq. (4.94) by Tanaka and Masunaga (2010) can also be used as an additional condition. The algorithm can be summarized as follows:

```
1:    if N_i then
2:        Return TRUE // particle i is a free-surface particle
3:    else if ⟨n⟩_i ≥ βn^0 or N_i ≥ β'N^0 then
4:        Return FALSE // particle i is NOT a free-surface particle
5:    else
6:        Calculate the moment matrix M_i via Eq. (4.97)
7:        Determine λ_i, the minimum eigenvalue of M_i
8:        if λ_i ≤ β_1 then
9:            Return TRUE // particle i is a free-surface particle
```

```
10:     else if λⱼ > β₂ then
11:        Return FALSE // particle i is NOT a free-surface particle
12:     else
13:        Calculate the unit normal vector nⱼ via Eq. (4.95)
14:        for j∈Λⱼ do
15:           if SCAN(rⱼ,rⱼ,nⱼ) = TRUE then
16:              Return FALSE // particle i is NOT a free-surface particle
17:           end if
18:        end for
19:        Return TRUE // particle i is a free-surface particle
20:     end if
21:  end if
```

The first examination in line 1 is to detect dispersed particles by the principle that for internal particles at least two neighbors per dimension should exist. The "if" statement in line 3 finds internal particles based on the criteria by Koshizuka and Oka (1996) and Tanaka and Masunaga (2010). In general, the majority of particles in system will be identified as either a free-surface particle or an internal one only by these rough determinations. Therefore, the calculation of the moment matrix and its eigenvalues in lines 6 and 7 as well as procedures for the scanning in line 13−18 can be skipped for them. For information on the thresholds, authors have obtained fair results with $\beta = \beta' = 0.9$, $\beta_1 = 0.2$, and $\beta_2 = 0.8$ for various problems. A calculation example with these settings is shown in Fig. 4.17E, where it should be noticed that mislabeling is much reduced compared to any of the conventional approaches.

Duan et al. (2017b) adopted an economical version of Marrone's method. They approximated the normal vector based on the barycenter of neighbor particles as

$$n_i = -\frac{b'_i}{|b'_i|} \tag{4.99}$$

$$b'_i = \sum_{j \in \Lambda_i} \frac{r_j - r_i}{|r_j - r_i|} \, w(|r_j - r_i|) \tag{4.100}$$

Notice here that the moment matrix is not included. This greatly reduces the number of arithmetic operations required to obtain the normal vector, while some drawbacks should be remarked. At first, from a mathematical point of view, the determination of normal vector should be more complete if one uses the inverse of the moment matrix, instead of the numerical barycenter. Moreover, the screening using the minimum eigenvalue λ cannot be performed since λ is not available here. This may increase the number of candidates to be examined through the high-cost scanning scheme. In fact, this version has an advantage particularly on ease of implementation in the calculation code, which is also an important feature for the practical use.

4.3.2 Pressure Calculation

Dirichlet boundary condition for pressure is imposed on the free-surface particles when solving the pressure Poisson equation. In the usual case, pressure value is fixed to be zero, which means the atmospheric pressure is assigned on the free surface. Thus, enforcement of the pressure boundary condition on free-surface particles itself is quite straightforward. Even for isolated particles, since the particles are identified to be on free surface based, for instance, on the reduction of neighbor particles, zero pressure will be assigned without special treatment. Such isolated particles perform a free-falling motion due to gravity, and will be incorporated back into the general pressure calculation after dropping onto the free surface and entrapped inside the liquid.

Treatment of pressure on the free-surface boundary is simple and convenient to deal with complicated motion of dispersed particles. However, additional numerical treatment is, in fact, almost necessary for the numerical stabilization. This is due to the so-called negative pressure problem. Here, "negative" means pressure being lower than that on the free surface. With the presence of negative pressure, impulsive forces acting between particles may take place. That, in some cases, remedies the particle spatial disorder, but in other cases, triggers severe numerical instability.

Negative pressure takes place either physically or numerically. Examples for the former case include flow behind solid object or rotating blade. On the other hand, the numerical negative pressure is caused mainly due to the particle deficiency. Recall that source term of the pressure Poisson equation contains the particle number density. Hence, particles that present near but not on free surface tend to have lower pressure than the free surface. Such numerical negative pressure may have a negative influence on the computation and thus should be removed.

One simple way to avoid this issue is to set an artificial lower bound for pressure field, so-called the zero limiter (Kondo and Koshizuka, 2011). After solving the pressure Poisson equation, negative pressure will be forced to be zero. By applying the zero limiter, simulation can be conducted stably. The technique has been adopted in most simulations. However, at the same time, it has significant side effects. Filtered pressure field not only removes the numerical negative pressure but also lose the information of physically meaningful negative pressure. As a consequence, unnatural void is frequently generated inside the liquid. This is seemingly occurrence of cavitation but, in fact, is of numerical error in most cases.

Apart from the negative pressure problem, particle clustering is another important issues that should be addressed. It is common to assign a constant pressure on free-surface particles for the Dirichlet boundary condition. Thus, repulsive forces are not properly applied between these particles even when they can get close to each other. This results in the formation of particle clusters on the free surface, and may introduce unphysical pressure oscillation.

Here we introduce various techniques that have been proposed to address these issues and to improve robustness and accuracy in the pressure calculation. Note that those numerical issues associated with free surface are basically not encountered when solved as a multiphase flow problem. Therefore, they can be considered as side effects of the vacuum approximation.

Lee et al. (2011) have proposed a collision model that can ease the particle clustering. The collision model adds repulsive forces between pairs of particles when they are closer than prescribed distance, so-called collision distance. The collision model modifies velocity by assuming two particles to be bounced each other. The velocity change due to the collision between particles i and j can be calculated by the following:

$$\Delta \boldsymbol{u}_{ij} = \begin{cases} \dfrac{1+b}{2} v_{ij} \dfrac{\boldsymbol{r}_j - \boldsymbol{r}_i}{|\boldsymbol{r}_j - \boldsymbol{r}_i|} & (|\boldsymbol{r}_j - \boldsymbol{r}_i| < al_0 \quad \text{and} \quad v_{ij} < 0) \\ 0 & (\text{otherwise}) \end{cases} \tag{4.101}$$

$$v_{ij} = \frac{(\boldsymbol{r}_j - \boldsymbol{r}_i) \cdot (\boldsymbol{u}_j - \boldsymbol{u}_i)}{|\boldsymbol{r}_j - \boldsymbol{r}_i|} \tag{4.102}$$

There are two tuning parameters: a and b. a is a coefficient for defining the collision distance. b is the coefficient of restitution. Both parameters are chosen in the range of $[0, 1]$. The parametric study by Lee et al. (2011) recommended to set $a = 0.2$ and $b = 0.9$. This model is effective to prevent free-surface particles from clustering, and also it can improve the overall uniformity of particle spacing if applied to internal particles. Thus, the collision model is commonly applied in the MPS method to stabilize the computation.

Before starting a discussion for the free-surface boundary conditions, let us overview important development toward the pressure calculation itself. Pressure field is determined as a solution of the pressure Poisson equation in the MPS method. The pressure Poisson equation was, in the original form,

$$\langle \nabla^2 P \rangle_i^{k+1} = \frac{\rho}{\Delta t^2} \frac{n^0 - \langle n \rangle_i^*}{n^0} \tag{4.103}$$

Here, the source term reflects the deviation of the particle number density. Thus, the corrections

$$u_i^{k+1} = u_i^* - \frac{\Delta t}{\rho} \langle \nabla P \rangle_i^{k+1} \tag{4.104}$$

$$r_i^{k+1} = r_i^* - \frac{\Delta t^2}{\rho} \langle \nabla P \rangle_i^{k+1} \tag{4.105}$$

can recover the particle number density to a constant (n^0). While this formulation is robust for conserving the fluid volume, it is well known that pressure tends to oscillate strongly in both spatial and time domain. On the other hand, in the mesh methods it is common to adopt the pressure Poisson equation in this form:

$$\langle \nabla^2 P \rangle_i^{k+1} = \frac{\rho}{\Delta t} \nabla \cdot u_i^* \tag{4.106}$$

Notice that the source term includes the velocity divergence instead of the particle number density. In fact, Eq. (4.106), hereafter referred to as the divergence-free form, can be used in the particle methods instead of Eq. (4.103), referred to as the density-invariant form. By doing this, pressure distribution becomes much smoother. However, difficulties are encountered because the fluid volume is not maintained constant with the divergence-free form. Aiming at improving the pressure distribution with good volume conservation, several approaches that combine both advantages have been developed. The approaches can be divided into two categories based on the strategy taken.

In the first category, the divergence-free form and the density-invariant form are both considered but treated separately. Sueyoshi and Naito (2004) have proposed an approach in which both Eq. (4.103) and (4.106) are solved separately. The solutions of pressure are stored in different memory, and corresponding velocity fields are also updated separately. Particle positions are updated based on the velocity field and obtained based on the density-invariant form. With this framework, smooth pressure distribution is obtained as a result of the divergence-free form calculation, and at the same time, fluid volume is conserved as the conventional density-invariant form. While their method should be effective to improve pressure distribution, it highly increases the computational complexity. On the other hand, Xu et al. (2009) have proposed a simplified approach. In the method, only the pressure Poisson equation of divergence-free form is solved. Instead of considering the density-invariant form, they adopted so-called the particle shifting technique to explicitly shift particle positions slightly to keep the particle spacing quasi-uniform. The position shift Δr_i can be calculated based fully on the structure of neighbor particles as follows:

$$\Delta r_i = -\alpha U_{\max} \Delta t \sum_{j \in \Lambda_i} \frac{r_j - r_i}{|r_j - r_i|^3} R_i^2 \tag{4.107}$$

$$R_i = \frac{1}{|\Lambda_i|} \sum_{j \in \Lambda_i} |r_j - r_i| \tag{4.108}$$

Here, $U_{\max}$ is maximum velocity magnitude, and α is a coefficient that adjusts the shifting distance. Generally, α is chosen in the range $[0.01, 0.1]$. The particle shifting was originally invented for problems without free surface. Recently, Lind et al. (2012) proposed a generalized version of the particle shifting, and Khayyer et al. (2017) proposed the optimized particle shifting scheme. They are applicable to free-surface flow problems.

The other category mixes the source term of both density-invariant and divergence-free forms, and considers them together by solving a single linear system. Tanaka and Masunaga (2010) have proposed

$$\langle \nabla^2 P \rangle_i^{k+1} = (1 - \gamma) \frac{\rho}{\Delta t} \nabla \cdot u_i^* + \gamma \frac{\rho}{\Delta t^2} \frac{n^0 - \langle n \rangle_i^*}{n^0} \tag{4.109}$$

where γ is a tuning parameter that takes a value from 0 to 1. Lee et al. (2011) suggest $0.01 < \gamma < 0.05$ from their numerical experiment. Pressure is mostly determined based on the velocity divergence term if γ is chosen small. Thus, the numerical oscillation can be well suppressed. Since the density variation term can capture the accumulation of volume change, it effectively improves the volume conservation property even with small γ in

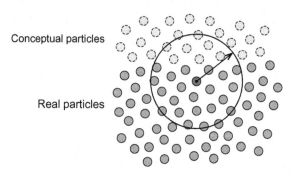

Figure 4.20 Conceptual particles.

most cases. Other formulae have been proposed by Kondo and Koshizuka (2011), Khayyer and Gotoh (2011), and Tamai and Koshizuka (2014).

Below we introduce techniques to improve pressure calculations for systems with free surface. As previously mentioned, in the presence of free-surface issues are often encountered. Two representatives are particle clustering on free surface, and the negative pressure problem. Particle clustering can be remedied by the collision model to some extent but not completely.

Chen et al. (2014) have invented a new computational scheme for enforcement of Dirichlet pressure boundary condition. Their method does not fix pressure on free surface, and even free-surface particles are not identified. Instead, it considers existence of virtual particles in gas phase, called conceptual particles as shown in Fig. 4.20. Conceptual particles have constant pressure denoted by P_{gas}. With the conceptual particles, one may define the modified particle number density by

$$\langle n' \rangle_i = \max[\langle n \rangle_i, \tilde{n}_i] \qquad (4.110)$$

$$\tilde{n}_i = n^0 + \sum_{j \in \Lambda_i} \max\left[w\left(\left|\mathbf{r}_j - \mathbf{r}_i\right|\right) - w(l_0), 0\right] \qquad (4.111)$$

With the introduction of conceptual particles and the modified particle number density, the pressure Poisson equation of Eq. (4.109) is modified as

$$\frac{2d}{\lambda n^0} \sum_{j \in \Lambda_i \cup \Lambda_i'} \left(P_j^{k+1} - P_i^{k+1}\right) w\left(\left|\mathbf{r}_j - \mathbf{r}_i\right|\right) = (1 - \gamma)\frac{\rho}{\Delta t}\nabla \cdot \mathbf{u}_i^* + \gamma\frac{\rho}{\Delta t^2}\frac{n^0 - \langle n' \rangle_i^*}{n^0} \qquad (4.112)$$

where Λ_i and Λ'_i indicate sets of neighbor real particles and conceptual particles, respectively. Now, pressure is given on conceptual particles to be P_{gas}, Eq. (4.112) can be rewritten as

$$\frac{2d}{\lambda n^0} \sum_{j \in \Lambda_i} \left(P_j^{k+1} - P_i^{k+1} \right) w\left(|\mathbf{r}_j - \mathbf{r}_i|\right) - \frac{2d}{\lambda n^0} \left[\langle n' \rangle_i - \langle n \rangle_i \right] P_i^{k+1}$$
$$= (1 - \gamma) \frac{\rho}{\Delta t} \nabla \cdot \mathbf{u}_i^* + \gamma \frac{\rho}{\Delta t^2} \frac{n^0 - \langle n' \rangle_i^*}{n^0} - \frac{2d}{\lambda n^0} \left[\langle n' \rangle_i - \langle n \rangle_i \right] P_{gas} \tag{4.113}$$

It is noticed that, in general, $\langle n' \rangle_i - \langle n \rangle_i > 0$ near free surface, and $\langle n' \rangle_i - \langle n \rangle_i = 0$ in inner region. In the latter case, Eq. (4.113) agrees with the original pressure Poisson equation, while in the former case, Dirichlet boundary condition is weakly imposed. Since pressure is not fixed at any particle, repulsive forces can take place between particles near the free surface. However, the conventional discretization scheme for the pressure gradient operator

$$\langle \nabla P \rangle_i = \frac{d}{n^0} \sum_{j \in \Lambda_i} \frac{P_j - \hat{P}_i}{|\mathbf{r}_j - \mathbf{r}_i|^2} \left(\mathbf{r}_j - \mathbf{r}_i \right) w\left(|\mathbf{r}_j - \mathbf{r}_i|\right) \tag{4.114}$$

$$\hat{P}_i = \min_{j \in \Lambda_i \cup \{i\}} P_j \tag{4.115}$$

is not suitable for preventing particle clustering. Another frequently used scheme is the conservative pressure gradient scheme proposed by Toyota et al. (2005).

$$\langle \nabla P \rangle_i = \frac{d}{n^0} \sum_{j \in \Lambda_i} \frac{P_j + P_i}{|\mathbf{r}_j - \mathbf{r}_i|^2} \left(\mathbf{r}_j - \mathbf{r}_i \right) w\left(|\mathbf{r}_j - \mathbf{r}_i|\right) \tag{4.116}$$

Compared to Eq. (4.114) that guarantees repulsive forces act between particles, which is important to stabilize the computation, the conservative scheme does not since attractive force happens when $P_j + P_i < 0$. However, this formulation renders better numerical stability in some cases. That is because Eq. (4.116) has larger repulsive force in magnitude for all pairs of particles when pressure is all positive. Chen et al. (2014) suggest to use the conservative scheme with conceptual particles. With such combination, their method significantly remedies particle clustering problem, and particle distributions with almost equal spacing can be obtained.

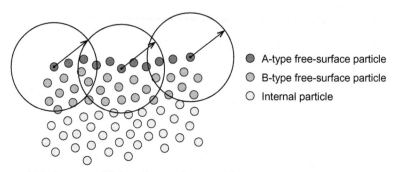

Figure 4.21 A-type and B-type free-surface particles.

It should be noted that in the source term of the pressure Poisson equation, the particle number density was replaced with the modified one. This guaranties the relation $n^0 - \langle n' \rangle_i^* < 0$ to be hold for all particles. Therefore, the density deviation term is always forced to be negative, and negative pressure will be eliminated. This can prevent particles to get too close, but it cannot prevent them to be sparse. In other words, particle spatial disorder such as occurrence of numerical cavity is not remedied with this method. To achieve such requirements, it is important to identify where negative pressure should be eliminated. Such identification has been given by Shibata et al. (2015a). They defined A-type and B-type free-surface particles (see Fig. 4.21). A-type is the same as the conventional free-surface particle. B-type is a neighbor of A-type free-surface particle but not A-type itself. By these definitions, particles not only on the free surface but also in the vicinity can be known. For those A-type and B-type particles, the particle number density tends to decrease due to the particle deficiency. Thus, negative pressure must be avoided because it results in severe numerical instability. Duan et al. (2017b) have conducted simulations of free-surface flow with such framework (Fig. 4.22). In their algorithm, the pressure Poisson equation and the pressure gradient scheme are switched for each particle depending on whether it is a free-surface particle (A-type or B-type), or it is an internal particle. Then only for the free-surface particles, the modified pressure Poisson equation of Eq. (4.113) and the conservative pressure gradient scheme of Eq. (4.116) are adopted. For the internal particles, the original pressure Poisson equation of Eq. (4.109) is applied. For the gradient scheme, the original scheme of Eq. (4.114) is applicable, but also another accurate scheme can also be adoptable, see for example, Eq. (33) in Duan et al. (2017b). In

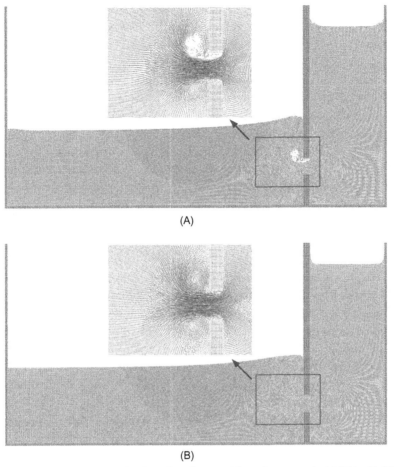

(A)

(B)

Figure 4.22 Simulation results for oil spilling problem (Duan et al., 2017b). (A) With the pressure limiter; (B) without the pressure limiter.

their calculation algorithm, negative pressure can be captured stably in internal region. Therefore, application of the zero limiter is not necessary. As a result, occurrence of the numerical cavity can be greatly suppressed, see their simulation results shown in Fig. 4.21.

4.4 INLET AND OUTLET BOUNDARY MODELING

Handling open systems with mass flux on the boundary is one of the key features for practical applications of computational fluid dynamics. Such configurations arise when the phenomenon of interest is restricted to only a small part of very large spatial domain over which the actual flow field extends. Examples include many important engineering applications such as fluid flow in a long pipe, turbomachinery, continuous chemical reactor, vehicle aerodynamics, and blood flow. It is infeasible to simulate these problems without the restriction of spatial domain. However, with the restricted computational domain, it is necessary to impose boundary conditions that approximate the interactions with the exterior. In particular, the boundary conditions typically accompany mass flux; in other words, flow into and out of the domain have to be dealt with. Such modeling has been well established and prevalently applied in the mesh methods. However, the particle methods encounter issues when dealing with these boundary conditions.

Since a particle method is a Lagrangian method, particles entering or leaving the local computational domain have to be considered for inlet and outlet boundaries. The inlet boundary is to represent inflow conditions where particles are actually injected into the domain with prescribed velocity profile or total mass flow rate. The outlet boundary defines an exit of flow where particles flow out through the boundary, on which usually pressure value is specified. These boundary conditions dealing with inflow and outflow are called open (or permeable) boundary conditions. The open boundaries are relatively rare in the MPS method as well as other particle methods.

To avoid difficulties associated with the particle deficiency problem, the open boundaries have frequently been modeled using the spatial periodicity, see Morris et al. (1997) for example. Periodic boundary conditions can represent repeated computational domains with ends connected to each other. Therefore, as illustrated in Fig. 4.23, the periodic boundary with a constant driving body force can be used to generate flow between two infinitely long parallel plates (so-called plane Poiseuille flow). However, the periodic boundary conditions can be applied only to such simple geometries. Total mass flow rate as well as the inlet velocity profile cannot be chosen arbitrary in an explicit way. Moreover, since particles flowing out the domain will be recycled as inflow particles, they can introduce numerical periodicity of certain frequency in flow field.

The basic framework for the numerical implementation of inlet and outlet boundaries in the MPS method has been provided by Koshizuka (2005). Fig. 4.24 shows a schematic diagram of his open boundary treatment. In the inlet boundary, particles with prescribed velocity must be

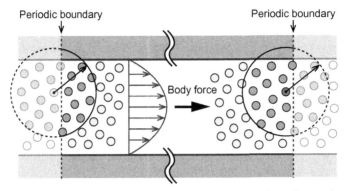

Figure 4.23 Periodic boundary condition representing open boundary with constant flow rate.

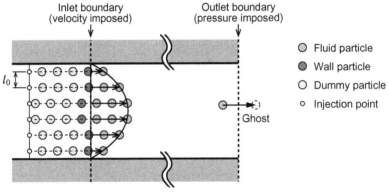

Figure 4.24 Schematic diagram of basic framework for the numerical implementation of inlet and outlet boundaries in the MPS method.

generated and inserted into the computational domain at a constant frequency. To this end, the wall particle representation (Section 4.2.1) is utilized.

At an initial state, wall particles and dummy particles are placed uniformly with an interparticle distance of l_0 outside the inlet boundary as if it is a solid wall boundary. However, we here use moving boundary particles to represent inflow conditions. The velocity of the boundary particles is given by the inlet boundary condition; and the particles move toward the boundary as time advances. The wall particles will become fluid particles when entered in the computational domain. At the same time, the innermost dummy particle on the same streamline will be converted to a wall particle, and a new dummy particle will be added at the corresponding injection point (see Fig. 4.24). By repeating this procedure, particles can be injected one after another. Note that for clarity, pressure of wall particle, but not dummy particle, is unknown variable and must be determined by solving the Poisson equation like as fluid particle. Therefore, the effective domain of wall particle must be fully covered by dummy particles. It means that the injection point should be placed at least $r_e + l_0$ (r_e is the effective radius) upstream from boundary.

Treatment for the outlet boundary conditions is straightforward compared to the inlet boundary. Outflow can be represented basically by removing particles that leave the computational domain. Generally, Dirichlet boundary conditions are given for pressure on the outlet boundary. Thus, pressure must be imposed on particles located near the boundary. Since particles flowing out the domain are deleted, particles adjacent to the boundary can be determined by means of the free-surface particle detection (Section 4.3.1). Therefore, free-surface particles within a certain distance are used as Dirichlet points of the outlet boundary condition for pressure calculation.

The total number of particles is variable over time in simulations with open boundary conditions. In such circumstances, it is important, in actual calculation code, to avoid frequent memory manipulations such as allocation and copying for reducing the computation time. This is particularly significant when dealing with a large number of particles. For these reasons, particles leaving through the outlet boundary are just labeled as "ghost" and are not actually deleted from memory. Here, ghosts are particles that are treated as if they no longer exist. Thus, no interaction will be calculated associated with them. Later, the memory space allocated for ghost particles will be recycled for inflow particles.

With the above-explained approach, one may encounter issues due to the mass flow rate inequality between inlet and outlet. This mass imbalance does not allow the simulations to conserve the total fluid volume (particle number) and thus to converge to a steady state solution. As reported by Shibata et al. (2015b), the conventional method fails to simulate flow past a cylinder accurately as shown in Fig. 4.25. The mass imbalance lets unnatural numerical cavities take place at the wake region, and thus, algorithmic sophistication is required to address this issue.

In this background, Shibata et al. (2015b) have developed an improved numerical treatment for the open boundary conditions. The key idea is to adjust the outflow velocity so that the total amount of mass in the computational domain is maintained constant. While the inlet boundary

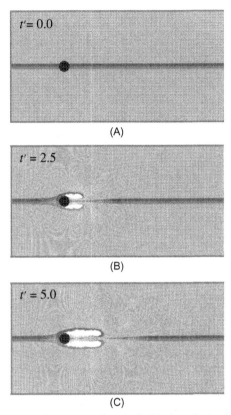

$t' = 0.0$

(A)

$t' = 2.5$

(B)

$t' = 5.0$

(C)

Figure 4.25 Development of a numerical cavity behind a cylinder (Shibata et al., 2015b). Fluid flows from left to right at Re = 100; left boundary is inlet; right boundary is outlet; upper and lower boundaries are solid walls.

condition is treated the same as the conventional approach, the outlet boundary condition is modified. At first, an outflow zone is defined outside the computational domain by extruding the outlet boundary in the downstream direction, as illustrated in Fig. 4.26. The length of the outflow zone was set to $2.1l_0$ in study of Shibata et al. (2015b). Particles within the outflow zone are used to impose Dirichlet pressure boundary condition and also to adjust the outflow rate. Dirichlet pressure is assigned to free-surface particles inside the outflow zone. Velocity in this domain is determined based on the following equations:

$$u_i = n \otimes n \ \tilde{u}_i + n \ \Delta u \qquad (4.117)$$

$$\Delta u = \frac{(N - N^0) l_0^d}{\alpha A \ \Delta t} \qquad (4.118)$$

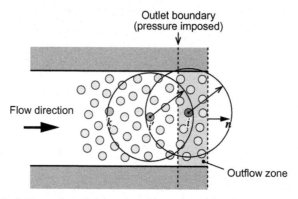

Figure 4.26 Definition of outflow zone and upstream particle.

Here, n is the outward (downstream direction) unit normal vector on the outlet boundary. Δu is a measure for the deviation of total fluid mass from the reference state. N is the current number of particles in the computational domain. N^0 is the reference value of N, which can be determined by the initial value of N. A is the area of the outlet boundary. α is a relaxation coefficient that is chosen close to unity. $\tilde{u}_i$ is the averaged upstream velocity measured around an upstream particle k. The upstream particle k is determined by the following procedure:

1. find a neighbor particle i' that fulfills

$$i' = \operatorname*{argmax}_{j \in \Lambda_i} \left[n \cdot \left(r_i - r_j \right) \right] \qquad (4.119)$$

2. determine the upstream particle k from the neighbors of i' based on

$$k = \operatorname*{argmax}_{j \in \Lambda_{i'}} \left[n \cdot \left(r_{i'} - r_j \right) \right] \qquad (4.120)$$

Through this procedure, a particle that is located about twice the effective radius upstream from the particle i in question should be chosen as the upstream particle k (see Fig. 4.26). The averaged upstream velocity $\tilde{u}_i$ is calculated by the simple weighted average around the particle k.

REFERENCES

Akimoto, H., 2013. Numerical simulation of the flow around a planing body by MPS method. Ocean Eng. 64, 72–79.

Asai M., Fujimoto K., Tanabe S., Beppu M., 2013. Slip and no-slip boundary treatment for particle simulation model with incompatible step-shaped boundaries by using a virtual marker. Transactions of the Japan Society for Computational Engineering and Science, Paper No. 20130011 (in Japanese).

Barker, D.J., Brito-Parada, P., Neethling, S.J., 2014. Application of B-splines and curved geometries to boundaries in SPH. Int. J. Numer. Methods Fluids 76, 51–68.

Belytschko, T., Krongauz, Y., Organ, D., Fleming, M., Krysl, P., 1996. Meshless methods: an overview and recent developments. Comput. Methods Appl. Mech. Eng. 139, 3–47.

Bierbrauer, F., Bollada, P.C., Phillips, T.N., 2009. A consistent reflected image particle approach to the treatment of boundary conditions in smoothed particle hydrodynamics. Comput. Methods Appl. Mech. Eng. 198, 3400–3410.

Blake, J.R., 1971. A note on the image system for a stokeslet in a no-slip boundary. Math. Proc. Cambridge Philos. Soc. 70, 303–310.

Børve S., 2011. Generalized ghost particle method for handling reflecting boundaries. In: Rung, T., Ulrich, C. (Eds.), Proceedings of the 6th SPHERIC International Workshop, Hamburg, Germany, pp. 313–320. http://www.tuhh.de/vss (ISBN 978-3-89220-658-3).

Chen, X., Xi, G., Sun, Z.-G., 2014. Improving stability of MPS method by a computational scheme based on conceptual particles. Comput. Methods Appl. Mech. Eng. 278, 254–271.

Chikazawa, Y., Koshizuka, S., Oka, Y., 1999. Numerical analysis of sloshing with large deformation of elastic walls and free surfaces using MPS method. Trans. Jpn Soc. Mech. Eng. Ser. B 65, 2954–2960 (in Japanese).

Colagrossi, A., Landrini, M., 2003. Numerical simulation of interfacial flows by smoothed particle hydrodynamics. J. Comput. Phys. 191, 448–475.

Cummins, S.J., Rudman, M., 1999. An SPH projection method. J. Comput. Phys. 152, 584–607.

Dilts, G.A., 2000. Moving least-squares particle hydrodynamics II: conservation and boundaries. Int. J. Numer. Methods. Eng. 48, 1503–1524.

Duan, G., Chen, B., Koshizuka, S., Xiang, H., 2017a. Stable multiphase moving particle semi-implicit method for incompressible interfacial flow. Comput. Methods Appl. Mech. Eng. 318, 636–666.

Duan, G., Chen, B., Zhang, X., Wang, Y., 2017b. A multiphase MPS solver for modeling multi-fluid interaction with free surface and its application in oil spill. Comput. Methods Appl. Mech. Eng. 320, 133–161.

Feldman, J., Bonet, J., 2007. Dynamic refinement and boundary contact forces in SPH with applications in fluid flow problems. Int. J. Numer. Methods Eng. 72, 295–324.

Ferrand, M., Laurence, D.R., Rogers, B.D., Violeau, D., Kassiotis, C., 2013. Unified semi-analytical wall boundary conditions for inviscid, laminar or turbulent flows in the meshless SPH method. Int. J. Numer. Methods Fluids 71, 446–472.

Haque, A., Dilts, G.A., 2007. Three-dimensional boundary detection for particle methods. J. Comput. Phys. 226, 1710–1730.

Harada T., Koshizuka S., Shimazaki K., 2008. Improvement of wall boundary calculation model for MPS method. Transactions of the Japan Society for Computational Engineering and Science, Paper No. 20080006 (in Japanese).

Hirt, C.W., Nichols, B.D., 1981. Volume of fluid (VOF) method for the dynamics of free boundaries. J. Comput. Phys. 39, 201–225.

Hirt, C.W., Amsden, A.A., Cook, J.L., 1974. An arbitrary Lagrangian–Eulerian computing method for all flow speeds. J. Comput. Phys. 14, 227–253.

Hosseini, S.M., Feng, J.J., 2011. Pressure boundary conditions for computing incompressible flows with SPH. J. Comput. Phys. 230, 7473–7487.

Huerta, A., Liu, W.K., 1988. Viscous flow with large free surface motion. Comput. Methods Appl. Mech. Eng. 69, 277–324.

Ikeda, H., Matsuura, F., Koshizuka, S., Oka, Y., 1998. Numerical analysis of fragmentation processes of liquid metal in vapor explosions using moving particle semi-implicit method. Trans. Jpn Soc. Mech. Eng. Ser. B 64, 2431–2437.

Khayyer, A., Gotoh, H., 2011. Enhancement of stability and accuracy of the moving particle semi-implicit method. J. Comput. Phys. 230, 3093–3118.

Khayyer, A., Gotoh, H., Shimizu, Y., 2017. Comparative study on accuracy and conservation properties of two particle regularization schemes and proposal of an optimized particle shifting scheme in ISPH context. J. Comput. Phys. 332, 236–256.

Kondo, M., Koshizuka, S., 2011. Improvement of stability in moving particle semi-implicit method. Int. J. Numer. Methods Fluids 65, 638–654.

Koshizuka S., "Ryushiho" Maruzen Shuppan (2005) ISBN-13: 978-4621075227. (in Japanese).

Koshizuka, S., Oka, Y., 1996. Moving-particle semi-implicit method for fragmentation of incompressible fluid. Nucl. Sci. Eng. 123, 421–434.

Koshizuka, S., Nobe, A., Oka, Y., 1998. Numerical analysis of breaking waves using the moving particle semi-implicit method. Int. J. Numer. Methods Fluids 26, 751–769.

Kulasegaram, S., Bonet, J., Lewis, R.W., Profit, M., 2004. A variational formulation based contact algorithm for rigid boundaries in two-dimensional SPH applications. Computat. Mech. 33, 316–325.

Lancaster, P., Salkauskas, K., 1981. Surfaces generated by moving least squares methods. Math. Comput. 37, 141–158.

Lee, B.-H., Park, J.-C., Kim, M.-H., Hwang, S.-C., 2011. Step-by-step improvement of MPS method in simulating violent free-surface motions and impact-loads. Comput. Methods Appl. Mech. Eng. 200, 1113–1125.

Libersky, L.D., Petschek, A.G., Carney, T.C., Hipp, J.R., Allahdadi, F.A., 1993. High strain Lagrangian hydrodynamics: a three-dimensional SPH code for dynamic material response. J. Comput. Phys. 109, 67–75.

Lind, S.J., Xu, R., Stansby, P.K., Rogers, B.D., 2012. Incompressible smoothed particle hydrodynamics for free-surface flows: a generalised diffusion-based algorithm for stability and validations for impulsive flows and propagating waves. J. Comput. Phys. 231, 1499–1523.

Liu, X., Xu, H., Shao, S., Lin, P., 2013. An improved incompressible SPH model for simulation of wave–structure interaction. Comput. Fluids 71, 113–123.

Marrone, S., Colagrossi, A., Le Touzé, D., Graziani, G., 2010. Fast free-surface detection and level-set function definition in SPH solvers. J. Comput. Phys. 229, 3652–3663.

Marrone, S., Antuono, M., Colagrossi, A., Colicchio, G., Le Touzé, D., Graziani, G., 2011. δ-SPH model for simulating violent impact flows. Comput. Methods Appl. Mech. Eng. 200, 1526–1542.

Matsunaga T., Shibata K., Murotani K., Koshizuka S., 2016. Fluid flow simulation using MPS method with mirror particle boundary representation. Transactions of the Japan Society for Computational Engineering and Science, Paper No. 20160002 (in Japanese).

Matsunaga T., Shibata K. and Koshizuka S., 2017. Boundary integral based approach for the wall representation in MPS method. Proceedings of the JSME 30th Computational Mechanics Division Conference, Osaka, Japan, 139 (in Japanese).

Mayrhofer, A., Ferrand, M., Kassiotis, C., Violeau, D., Morel, F.-X., 2015. Unified semi-analytical wall boundary conditions in SPH: analytical extension to 3-D. Numer. Algorith. 68, 15–34.

Mitsume, N., Yoshimura, S., Murotani, K., Yamada, T., 2015. Explicitly represented polygon wall boundary model for the explicit MPS method. Computat. Particle Mech. 2, 73–89.

Morris, J.P., Fox, P.J., Zhu, Y., 1997. Modeling low Reynolds number incompressible flows using SPH. J. Comput. Phys. 136, 214–226.

Peskin, C.S., 1972. Flow patterns around heart valves: a numerical method. J. Comput. Phys. 10, 252−271.

Reshetouski, I., Ihrke, I., 2013. Mirrors in computer graphics, computer vision and time-of-flight imaging. In: Grzegorzek, M., Theobalt, C., Koch, R., Kolb, A. (Eds.), Time-of-Flight and Depth Imaging: Sensors, Algorithms and Applications. Springer, Berlin Heidelberg, pp. 77−104.

Rider, W.J., Kothe, D.B., 1998. Reconstructing volume tracking. J. Comput. Phys. 141, 112−152.

Ryskin, G., Leal, L.G., 1983. Orthogonal mapping. J. Comput. Phys. 50, 71−100.

Shibata, K., Masaie, I., Kondo, M., Murotani, K., Koshizuka, S., 2015a. Improved pressure calculation for the moving particle semi-implicit method. Computat. Particle Mech. 2, 91−108.

Shibata, K., Koshizuka, S., Murotani, K., Sakai, M., Masaie, I., 2015b. Boundary conditions for simulating Karman vortices using the MPS method. J. Adv. Simulat. Sci. Eng. 2, 235−254.

Sueyoshi, M., Naito, S., 2004. An improvement on pressure calculation scheme of MPS method. J. Kansai Soc. Naval Archi. 2004, 53−60 (in Japanese).

Sussman, M., 2003. A second order coupled level set and volume-of-fluid method for computing growth and collapse of vapor bubbles. J. Comput. Phys. 187, 110−136.

Szewc, K., Pozorski, J., Minier, J.-P., 2012. Analysis of the incompressibility constraint in the smoothed particle hydrodynamics method. Int. J. Numer. Methods Eng. 92, 343−369.

Tamai, T., Koshizuka, S., 2014. Least squares moving particle semi-implicit method. Computat. Particle Mech. 1, 277−305.

Tanaka, M., Masunaga, T., 2010. Stabilization and smoothing of pressure in MPS method by quasi-compressibility. J. Comput. Phys. 229, 4279−4290.

Toyota E., Akimoto H., Kubo S., 2005. A particle method with variable spatial resolution for incompressible flows. In: Proceedings of the 19th Japan Society of Fluid Mechanics, vol. 9-2.

Unverdi, S.O., Tryggvason, G., 1992a. A front-tracking method for viscous, incompressible, multi-fluid flows. J. Comput. Phys. 100, 25−37.

Unverdi, S.O., Tryggvason, G., 1992b. Computations of multi-fluid flows. Phys. D Nonlinear Phenom. 60, 70−83.

Xu, R., Stansby, P., Laurence, D., 2009. Accuracy and stability in incompressible SPH (ISPH) based on the projection method and a new approach. J. Comput. Phys. 228, 6703−6725.

Yildiz, M., Rook, R.A., Suleman, A., 2009. SPH with the multiple boundary tangent method. Int. J. Numer. Methods. Eng. 77, 1416−1438.

CHAPTER 5

Surface Tension Models in Particle Methods

Abstract

Particle methods have an advantage on the calculations having free surface flow because of its ability to track the motion directly with the particle movement. The free surface motion is highly dependent on the surface tension as well as pressure, viscosity, and gravity. In this chapter, the models to calculate surface tension in particle methods are shown. In specific, the model based on continuum surface force (CSF) and the model using pairwise potential force are presented. The further development and its application of the models are also described.

Keywords: Surface tension; continuum surface force; pairwise potential; wettability; microscale problem; breakup behavior; bubble calculation

Contents

5.1 Surface Tension Calculation Using CSF Equation		218
5.1.1 CSF-Based Model Proposed by Nomura et al. (2001)		218
5.1.2 Other Surface Tension Models Based on CSF Equation		222
5.2 Surface Tension Calculation Based on a Pairwise Potential		223
5.2.1 Potential-Based Model Proposed by Kondo et al. (2007a,b)		224
5.2.2 Further Improvement of the Potential-Based Approach		227
5.2.3 Wettability Calculation in the Potential Model		227
5.3 Applications of the Surface Tension Models Using the MPS Method		229
References		231

To calculate the breakup behavior or the flows in microscale, surface tension is to be considered. In particle methods, there are mostly two types of surface tension models: one is CSF (continuum surface force) model, which is also applied in mesh methods as in VOF (volume of fluid). The other is potential model, in which the pairwise attractive force is applied to the particles. Both the types of surface tension model have been developed also in the framework of MPS (moving particle semi-implicit) method. Regardless of the surface tension model, the surface tension force is usually calculated in the explicit step of the semi-implicit algorithm. Fig. 5.1 shows the flow chart of the surface tension calculation.

Moving Particle Semi-implicit Method
DOI: https://doi.org/10.1016/B978-0-12-812779-7.00005-9

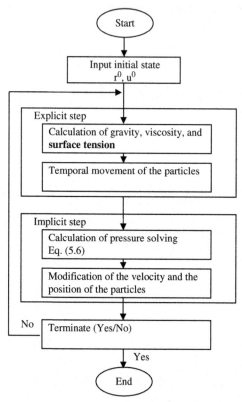

Figure 5.1 Surface tension calculation in the MPS algorithm. *Source: Kondo, M., Koshizuka, S., Suzuki, K., Takimoto, M. 2007b. Surface tension model using inter-particle force in particle method. Proceedings of the FEDSM2007, 5th Joint ASME/JSME Fluids Engineering Conference, FEDSM2007-37192.*

5.1 SURFACE TENSION CALCULATION USING CSF EQUATION

5.1.1 CSF-Based Model Proposed by Nomura et al. (2001)

In this section, the surface tension model proposed by Nomura et al. (2001) is shown. It is the CSF model firstly proposed in the MPS framework.

The governing equation with the CSF can be expressed as:

$$\rho \frac{D\mathbf{u}}{Dt} = -\nabla P + \mu \nabla^2 \mathbf{u} + \rho \mathbf{g} + \sigma \kappa \delta \mathbf{n} \tag{5.1}$$

where ρ, $\mathbf{u}$, P, μ, $\mathbf{g}$, σ, κ, δ, and $\mathbf{n}$ are fluid density, fluid velocity, pressure, viscosity, gravity, curvature of the surface, delta function on the surface, and normal vector of the surface, respectively. The last term on the right-hand side is the surface tension term, while the first, the second, and the third terms are pressure term, viscosity term, and gravity term, respectively. The surface tension force is limited only to the surface by the delta function, therefore the force is called CSF (Brackbill and Zemach, 1992). To model surface tension in a particle method like MPS method using this CSF equation, the surface tension term has to be discretized into particle interaction force, where discretization of the curvature κ, the normal vector $\mathbf{n}$, and delta function δ are to be considered.

To calculate surface tension based on the CSF model, the surface detection is needed so as to limit the force only to the surface particles. However, an additional procedure for the surface detection is not needed because the calculation for surface detection is originally included in the algorithm of MPS method. In the MPS method, the surface detection is conducted using the parameter called particle number density, which is defined as a summation of the weight function as:

$$n_i = \sum_{j \neq i} w(|\mathbf{r}_j - \mathbf{r}_i|) \tag{5.2}$$

where the weight function $w(r)$ is defined as:

$$w(r) = \begin{cases} \dfrac{r_e}{r} - 1 & (r < r_e) \\ 0 & (r \geq r_e) \end{cases} \tag{5.3}$$

The particles having the particle number density less than the threshold are judged as surface particles. Specifically, the particles satisfying the condition

$$n_i < \beta n_0$$

are regarded as surface, where n_0 is the constant representing the particle number density in the fluid bulk. In MPS, $\beta = 0.97$ is empirically used for the threshold value. The CSF-based surface tension force is only given to the particles assumed as surface particles. After the surface detection, the particles located in some finite area close to the surface are judged as surface particles. The depth of the surface area d_{st} is depending on the threshold β. Empirically, the depth d_{st} is 1.55 times larger than the particle distance l_0 ($d_{st} = 1.55 l_0$) when we adopt the threshold $\beta = 0.97$. Here, the parameter $d_{st}/l_0 = 1.55$ means how many layers of particle are in the surface area (Fig. 5.2). In discretizing the delta function, we have to be careful of the

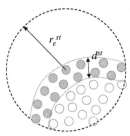

Figure 5.2 Thickness of the surface area. *Source: Nomura, K., Koshizuka, S., Oka, Y., Obata, H., 2001. Numerical analysis of droplet breakup behavior using particle method. J. Nucl. Sci. Technol. 38 (12), 1057–1064.*

feature that integration of the delta function is unity. To keep this important feature, the delta function is replaced by $l_0/d_{st} = 1/1.55$.

To estimate the curvature, another particle number density using different weight function is calculated. In short, the curvature is evaluated by comparing the particle number density at each surface particle and the base particle number density at the flat surface. When the surface is concave, the particle number density will be smaller than the base value, and when the surface is convex, it will be larger than the base value. To evaluate the surface curvature in this manner, the particles outer than itself should be ignored in calculating the particle number density. The calculation is done in two steps:

First, precalculate the particle number density n_i^{st1} as:

$$n_i^{st1} = \sum_{j \neq i} w^{st1}(|\mathbf{r}_j - \mathbf{r}_i|) \qquad (5.4)$$

where the weight function w^{st1} is defined as:

$$w^{st1}(r) = \begin{cases} 1 & (r < r_e^{st}) \\ 0 & (r \geq r_e^{st}) \end{cases} \qquad (5.5)$$

In the surface tension calculation, the interaction radius rest is set larger than the radius for other calculation. For example, it is set $r_e^{st} = 3.1 l_0$. The particle number density n^{st2}, which is used to estimate curvature is then calculated as:

$$n_i^{st1} = \sum_{j \neq i} w^{st1}(|\mathbf{r}_j - \mathbf{r}_i|) \qquad (5.6)$$

where the weight function w^{st2} is defined as:

$$w^{st2}(r) = \begin{cases} 1 & (r < r_e^{st} \text{ and } n_j^{st1} > n_i^{st1}) \\ 0 & (r \geq r_e^{st}) \end{cases} \tag{5.7}$$

In the calculation of n^{st2}, particles having smaller n^{st1} than itself are regarded as outer particles and skipped in summing up the weight function. The curvature is estimated as:

$$2\theta_i = \frac{n_i^{st2}}{n_0^{st}}\pi \tag{5.8}$$

and

$$\kappa_i = \frac{1}{R_i} = \frac{2\cos\theta_i}{r_e^{st}} \tag{5.9}$$

where n_0^{st} is the base particle number density, which is estimated at the surface on the flat surface.

Another parameter to be estimated for CSF calculation is the normal vector. In the CSF model proposed by Nomura et al. (2001), the gradient of the particle number density is estimated using the particle distance l_0 and the particle number density at the four shifted positions from particle i, which are denoted by n^{+x}, n^{-x}, n^{+y}, n^{-y}, respectively. The formulations for the gradient vector $\mathbf{a}$ is:

$$\mathbf{a}_i = \frac{n_i^{+x} - n_i^{-x}}{2l_0}\mathbf{n}_x + \frac{n_i^{+y} - n_i^{-y}}{2l_0}\mathbf{n}_y \tag{5.10}$$

where n^x and n^y are the unit vector to x-direction and y-direction, respectively. The normal vector is estimated by normalizing the gradient vector $\mathbf{a}$ as:

$$\mathbf{n}_i = \frac{\mathbf{a}_i}{|\mathbf{a}_i|} \tag{5.11}$$

Using these estimations, which are delta function $l_0/d^{st} = 1/1.55$, the curvature in Eq. (5.9), and the normal vector in Eq. (5.11), the CSF force is formulated as:

$$\sigma\kappa\delta\mathbf{n} = \sigma\kappa_i\frac{l_0}{d^{st}}\mathbf{n}_i \tag{5.12}$$

The formulation shown in Eq. (5.12) is the discretized CSF in MPS. However, the force is sometimes calculated unrealistically large because of the bad estimation of the model. Since it is not good for the stable calculation, special smoothing is given to the force before accelerating the particles in practice.

5.1.2 Other Surface Tension Models Based on CSF Equation

Other than the CSF calculation proposed by Nomura et al., the CSF-based surface tension calculations are proposed by several researchers in the framework of MPS method. To use CSF equation, the delta function, the normal vector, and the curvature are usually be formulated in some way. In fact, several different alternatives (Ishii et al., 2006; Ikejiri et al., 2007; Ichikawa and Labrosse, 2010; Duan et al., 2015) are proposed for each.

One of the alternatives for the estimation of the normal vector is using color function, which is firstly proposed by Ishii et al. (2006), and similar approaches are taken by several researchers. The color function c is the same in the same phase, and it takes other value in the other phase. The normal vector estimation got simple by using the color function c. For example, the normal vector multiplied by the delta function can be written as:

$$\delta\mathbf{n} = \frac{\nabla c}{|c|} \tag{5.13}$$

where $|c|$ is the difference of the color function between two different phases. In the single phase calculation, the color function c takes 1 at the particle in fluid phase and the value will be automatically zero in the area where no particle exists. The partial differential in Eq. (5.13) is to be discretized using the particle interaction models.

In most of the CSF calculation using color function, the following curvature estimation is done using the normal vector. The curvature can be written as a divergence of the normal vector as:

$$\kappa = \nabla \cdot \mathbf{n} \tag{5.14}$$

This approach is straightforward and can achieve the better estimation of the curvature. In fact, Duan et al. (2015) proposed the formulation based on the color function to achieve higher order accuracy, and it is pointed out that the curvature estimation using the particle number density (Nomura et al., 2001) will be bad when the curvature radius is very small compared to the effective radius (Shibata et al., 2004). However, the numerical stability in estimating curvature using particle number density is better than that in using divergence of the normal vector. When we adopt the CSF approaches, it suffers surface scattering because the local fluctuation cannot be recovered easily. This tendency is stronger when we adopt the curvature estimation using the divergence of the normal vector. When we approximate Eqs. (5.13) and (5.14), two particle interaction models, which are gradient model and divergence model, are used. Therefore, the particle locations in the range of $2r_e$ are used. On the other hand, when we use the particle number density, only the particle locations within r_e are used. Therefore, it is considered that the local particle distribution is more likely to be reflected to the curvature estimation, and it may yield some force to recover the local fluctuation on the surface. This might be a possible reason for the better numerical stability in the CSF approach using the particle number density (Nomura et al., 2001).

In most of the CSF-based approach, the force acting on the particles is directly calculated. On the other hand, Wada et al. (2015) have applied the other approach, in which the force is not calculated directly. Instead, the pressure rise with respect to the surface tension is given as a boundary pressure in the pressure Poisson's equation. When the surface is curved, there will be a pressure gap, and the gap can be expressed as:

$$\Delta P = \sigma \kappa \tag{5.15}$$

following the CSF equation. This approach can simplify the calculation to some extent; however, it also suffers the complexity in the evaluation of the normal vector and the curvature.

5.2 SURFACE TENSION CALCULATION BASED ON A PAIRWISE POTENTIAL

In the CSF model, the governing equations including CSF has to be discretized, which needs complex formulations to treat the partial differentials. On the other hand, there is another type of surface tension model, where the surface tension can be expressed only by introducing a pairwise potential. The potential force is similar to the molecular force, which is long-range attractive and short-range repulsive. Compared to CSF approach, the formulation is much simpler, and surface scattering is not likely to occur. The potential-based approach in the MPS framework is firstly suggested by Shirakawa et al. (2001a,b) and the relation between the surface tension and the pairwise potential is shown by Kondo et al. (2007a,b).

5.2.1 Potential-Based Model Proposed by Kondo et al. (2007a,b)

In this section, the surface tension calculation proposed by Kondo et al. (2007a,b) is presented specifically. A potential which derives short-range repulsive and long-range attractive force is formulated as:

$$P(r) = Cp(r)$$

$$p(r) = \begin{cases} -\dfrac{1}{3}\left(r - \dfrac{3}{2}l_0 + \dfrac{1}{2}r_e\right)(r-r_e)^2 & (r < r_e) \\ \\ 0 & (r \geq r_e) \end{cases} \qquad (5.16)$$

where C is for calibration, and r, r_e, l_0 are the particle distance between particle pair, the effective radius in surface tension calculation, and the initial particle spacing, respectively. The potential shown in Eq. (5.16) is just a polynomial function that it can easily be differentiated and be calculated. The shape of the potential is shown in Fig. 5.3. The pairwise potential will be minimum at $r = l_0$. The total potential E_{st} with respect to this force is:

$$E_{st} = \sum_{i,j} P(r_{ij}) \qquad (5.17)$$

where r_{ij} is given by:

$$r_{ij} = |\mathbf{r}_j - \mathbf{r}_i| \qquad (5.18)$$

By differentiating the potential with respect to particle coordinate $\mathbf{r}_i$, the corresponding interaction force acting at particle i can be derived as:

$$\mathbf{f}_i^{st} = -\frac{\partial E_{st}}{\partial \mathbf{r}_i} = \sum_{j \neq i} \frac{\partial P(r_{ij})}{\partial r_{ij}} \frac{\mathbf{r}_{ij}}{r_{ij}} = -\sum_{j \neq i} C(r_{ij} - l_0)(r_{ij} - r_e)\frac{\mathbf{r}_{ij}}{r_{ij}} \qquad (5.19)$$

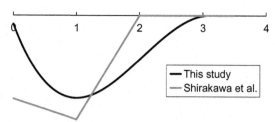

Figure 5.3 Shape of the potential function. *Source: Kondo, M., Koshizuka, S., Suzuki, K., Takimoto, M. 2007b. Surface tension model using inter-particle force in particle method. Proceedings of the FEDSM2007, 5th Joint ASME/JSME Fluids Engineering Conference, FEDSM2007-37192.*

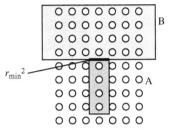

Figure 5.4 Potential energy against surface creation. *Source: Kondo, M., Koshizuka, S., Suzuki, K., Takimoto, M. 2007b. Surface tension model using inter-particle force in particle method. Proceedings of the FEDSM2007, 5th Joint ASME/JSME Fluids Engineering Conference, FEDSM2007-37192.*

where r_{ij} is the relative vector from particle i to j defined as:

$$\mathbf{r}_{ij} = \mathbf{r}_j - \mathbf{r}_i \tag{5.20}$$

The arbitrary parameter C included in Eq. (5.16) should be determined in some way before the calculation. Since the surface tension is equivalent to the surface energy, the coefficient C can be determined by relating to the surface energy. The surface energy can be estimated using Fig. 5.4. When the particles in area A and particles in area B are detached, two of the surface area having S will be created. Since the surface energy is the energy given against the pairwise potential in creating the surface, the energy can be estimated by adding up the potentials between area A and area B. This is expressed in the relation as:

$$2\sigma l_0^2 = \sum_{i \in A, j \in B} P(r_{ij}) \tag{5.21}$$

Using this relation, the surface tension (surface energy) will be estimated as:

$$\sigma = \frac{\sum_{i \in A, j \in B} P(r_{ij})}{2l_0^2} \tag{5.22}$$

The calculation example using the potential-based surface tension model (Kondo et al., 2007a,b) is shown in Fig. 5.5. It was a two-dimensional calculation. The squared shaped droplet got round due to the surface tension and oscillated without surface scattering. Although the oscillation decay was fast, the oscillation period roughly agreed with the theoretical value.

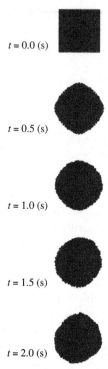

$t = 0.0$ (s)

$t = 0.5$ (s)

$t = 1.0$ (s)

$t = 1.5$ (s)

$t = 2.0$ (s)

Figure 5.5 Droplet oscillation using the potential model. *Source: Kondo, M., Koshizuka, S., Suzuki, K., Takimoto, M. 2007b. Surface tension model using inter-particle force in particle method. Proceedings of the FEDSM2007, 5th Joint ASME/JSME Fluids Engineering Conference, FEDSM2007-37192.*

In the potential-based surface tension calculation, only to be introduced to particle methods are the force in Eq. (5.19). This simplicity is one of the merits using the potential-based surface tension model. Another merit of potential model is numerical stability. Because the force is derived from the potential in the manner similar to the microscale method such as MD (molecular dynamics), the mechanical energy conservation with respect to this term is assured. It is very important to conduct stable calculation. Since the force acting on the particles are calculated only depending on the distance between neighbor particles, the local fluctuation of the particle distribution can be directly reflected to the force. This is helpful in avoiding the surface scattering which is one of the issues to be overcome in CSF approach (Ishii et al., 2006).

However, in the potential-based surface tension model, the force acts not only on the surface but also in the fluid bulk. This is different from the

CSF model, which is shown in the previous section. In the governing equations with surface tension (Eq. 5.1), the surface tension is limited to the surface by the delta function. Therefore, the force in the potential model is not equivalent to the surface tension term in the governing equation (Eq. 5.1). The excessive force of the potential model is attractive force, and it also acts inside the fluid bulk. To cancel the bulk force, the additional repulsive force is needed. However, it is not necessary to explicitly add some procedure because the pressure term can yield the repulsive force by calculating higher pressure to push against the excessive bulk force.

5.2.2 Further Improvement of the Potential-Based Approach

Since the surface tension model based on the pairwise potential (Kondo et al., 2007a,b) has better numerical stability than the approach using CSF equation, it can be applied to the industrial problems with complex geometry and multiphysics problems. However, the pairwise force acting inside the fluid bulk makes it difficult to conduct quantitative evaluation accurately. Therefore, several researchers (Ishii et al., 2012) have tried to subtract the inner force from the pairwise potential model. Ishii et al., 2012 canceled the pressure rise due to the force by subtracting the pressure rise on the flat surface. He converted the potential force to the pressure as:

$$P_n = \frac{\mathbf{f}_i \cdot \mathbf{n}_i}{A} \tag{5.23}$$

where $\mathbf{n}_i$ and A are the normal vectors at particle i and the surface area around particle i. After evaluating the pressure rise on the flat surface, the force acting on the surface particles were calculated as:

$$\overrightarrow{F_\nu} = \frac{P_n - P_n|_{flat}}{h} d\overrightarrow{n} \tag{5.24}$$

With this procedure, the pressure rise due to the bulk force can be canceled. However, the algorithm will be complex because it requires the additional procedure for the surface detection, normal vector evaluation, and the pressure evaluation.

5.2.3 Wettability Calculation in the Potential Model

In the potential-based surface tension calculation, the pairwise potential force is given between the fluid particles. The similar force can also be given between fluid particles and solid particles, which is useful to express the wettability of the fluid to the wall. To express the wettability, a

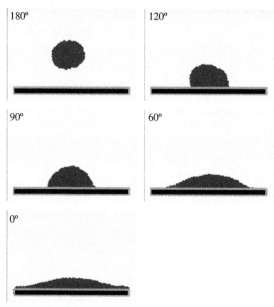

Figure 5.6 Static contact angle calculation using the potential model. *Source: Kondo, M., Koshizuka, S., Suzuki, K., Takimoto, M. 2007b. Surface tension model using inter-particle force in particle method. Proceedings of the FEDSM2007, 5th Joint ASME/JSME Fluids Engineering Conference, FEDSM2007-37192.*

parameter called contact angle is used. In the calculation using the CSF-based surface tension model, the movement of the particles close to the wall is forced to form the contact angle (Harada et al., 2005). Besides in the calculation using the potential-based approach, the contact angle can be simulated via the balance of the interaction forces, similar to the atomistic scale mechanism through which the contact angle emerges. Based on the Young's equation, the relation between the potential force and the contact angle θ can be derived as:

$$C_{fs} = \frac{C_f}{2}(1 + \cos\theta) \tag{5.25}$$

where C_{fs} and C_f are the coefficients for the fluid—solid potential and the fluid—fluid potential, respectively.

The static contact angle calculation using the potential-based approach (Kondo et al., 2007a,b) is shown in Fig. 5.6. Although the quantitative agreement was not so good in small contact angle, the tendency of the wettability can be expressed well.

5.3 APPLICATIONS OF THE SURFACE TENSION MODELS USING THE MPS METHOD

The model based on the CSF is firstly proposed by Nomura et al. (2001) and which is widely used in the MPS framework. Since the MPS method is proposed in the nuclear engineering field, the CSF-based model was firstly applied to analyze the breakup behavior of the corium in a severe accident (Nomura et al., 2001). Duan et al. (2003) calculated the droplet oscillation to verify the method, and they also applied it to the droplet breakup close to the critical We number. Besides, Shibata et al. (2004) calculated the jet breakup behavior. Since the jet breakup behaviors in a nuclear severe accident is multiphysics phenomena, they are not captured only with the surface tension calculation. Therefore the surface tension model was combined with other physical models. One of the examples is flashing calculation conducted by Duan et al. (2006), where the boiling model is applied together with the CSF model. The CSF model was applied not only to the nuclear engineering field but also to the other industrial field. Gotoh et al. (2003) applied the method in the civil engineering field and simulated free surface motion of the ocean wave. Tanaka et al. (2004) calculated gravitational oscillation in a perpendicular pipe in agricultural engineering field. Harada et al. (2005, 2006) applied the method to calculate the droplet behavior in a micro device, to clarify the mechanism of droplet generation, and similar study was also conducted by Rong and Chen (2010). Since the surface tension is largely related to the phase separation, it is used in the multiphase calculation. Park and Jeun (2011) calculated water droplet impingement, and Kim et al. (2014) calculated multiphase sloshing to clarify the mechanism of phase separation process. However, the curvature approximation proposed by Nomura et al. (2001) does not have so good accuracy.

To obtain the higher accuracy in the CSF approach, the alternative curvature approximations based on the color function, which is the label number to each phase, were proposed by some researchers (Ishii et al., 2006; Ikejiri et al., 2007; Ichikawa and Labrosse, 2010; Duan et al., 2015). They are mainly applied to the multiphase flow where the color function approach is suitable. In fact, Ishii et al. (2006) has applied the method to the breakup phenomena of the spray where the gas phase is also considered. Ikejiri et al. (2007) applied a CSF-based surface tension model for a single rising bubble in the hybrid particle mesh method. Ichikawa and Labrosse (2010) proposed a CSF-based surface tension model, where the normal vector was approximated via smooth particle approach while the curvature was approximated using the Laplacian model in the MPS framework. Duan et al. (2015) developed the higher order formulation based on the CSF equation, and the basic deformation of a single droplet is calculated for verification. Shakibaeinia (2015) applied a CSF-based approach in the air—water interaction calculation with an LES (large eddy simulation) turbulence model. Wada et al. (2015, 2016) calculated an ultrasonically levitated droplet using a CSF approach. It seems that the CSF approaches using color function are suitable for multiphase problems, where capturing the surface in high accuracy is required.

On the other hand, the potential-based models were applied to the relatively complex problems because of the simplicity and the stability. Shirakawa (2000), Shirakawa et al. (2001a,b, 2002, 2003), and Zhang et al. (2004) firstly proposed the method to calculate surface tension based on the pairwise potential. They applied the method to the two—phase flows going through the complex geometry in the nuclear reactors. The idea of the potential model was inherited by Kondo et al. (2007a,b), and the relation between the surface tension coefficient and the pairwise force was shown. In the model proposed by Kondo et al., a polynomial pairwise potential was used. The potential model has widely being applied in the complex industrial problems, and its improvement was proposed by several researchers Ishii and Sugii (2012), Inagaki (2015). Yoneda and Takimoto (2007) applied the potential model to simulate the flow through the complex geometry of GDL (gas diffusion layer) in an FC (fuel cell). Ido et al. (2011) applied the method to calculate the liquid bridge in the granular material and the sliding droplet on the slope, where the contact angle to the solid object plays an important role. Michiwaki et al. (2014) adopted the potential-based surface tension model to the swallowing simulation, where the surface tension of the food texture was taken into consideration. Nodomi (2015) conducted a three-dimensional calculation to simulate the complex flow characteristics of the resin using the surface tension model. Inagaki (2015) applied the model to the melting/solidification simulation, whose aim is to know the phenomena in a nuclear severe accident. Hattori et al. (2016) proposed a new contact angle calculation methodology and calculated the capillary bridge, which is to clarify the behavior of the condensate in a car air conditioner. Saso and Mouri (2017) used the potential-based surface tension model in combination with melting model and Marangoni convection model for welding simulation. It is considered that the potential models are suitable for the problems having high complexity where the numerical stability is important. In fact, they were applied to the problems having complex geometry and the multiphysics problems as shown earlier.

REFERENCES

Brackbill, J.U., Zemach, C., 1992. A continuum method for modelling surface tension. J. Comput. Phys. 100, 335−354.

Duan, R.-Q., Koshizuka, S., Oka, Y., 2003. Two-dimensional simulation of drop deformation and breakup at around the critical Weber number. Nucl. Eng. Design 225, 37−48. (臨界We数).

Duan, R.-Q., Jiang, S.-Y., Koshizuka, S., Oka, Y., Yamaguchi, A., 2006. Direct simulation of flashing liquid jets using the MPS method. Int. J. Heat and Mass Transfer 49, 402.

Gotoh, H., Ikari, H., Yagi, T., Sakai, T., 2003. Proc. Coastal Eng. 50, 21−25 (in Japanese).

Harada, T., Suzuki, Y., Koshizuka, S., Arakawa, T., Shoji, S., 2005. Simulation of droplet generation in micro flow using MPS method. Transact. Jpn Soc. Mech. Eng. Series B 79 (808), 2637−2641 (in Japanese).

Harada, T., Suzuki, Y., Koshizuka, S., Arakawa, T., Shoji, S., 2006. Simulation of droplet generation in micro flow using MPS method. JSME Int. J. Series B 49 (3), 731−736.

Kim, K.S., Kim, M.H., Park, J.-C., 2014. Development of moving particle simulation method for multi-liquid-layer sloshing. Math. Problems Eng. Article ID 350165, (2014).

Nomura, K., Koshizuka, S., Oka, Y., Obata, H., 2001. Numerical analysis of droplet breakup behavior using particle method. J. Nucl. Sci. Technol. 38 (12), 1057−1064.

Park, S., Jeun, G., 2011. Calculation of water droplet impingement using the coupled method of rigid body dynamics and the moving particle semi-implicit method. J. Mech. Sci. Technol. 25 (11), 2787−2794.

Rong, S., Chen, B., 2010. Numerical simulation of Taylor bubble formation in microchannel by MPS method. Microgravity Sci. Technol. 22, 321−327.

Shibata, K., Koshizuka, S., Oka, Y., 2004. Numerical analysis of jet breakup behavior using particle method. Transact. JSCES Paper No. 20040013. (in Japanese).

Tanaka, Y., Mukai, A., Taruya, H., 2004. The flow analysis in perpendicularly made pipe. Tech. Rep. Natl. Inst. Rural Eng. 202, 101−111 (in Japanese).

Duan, G., Koshizuka, S., Chen, B., 2015. A contoured continuum surface force model for particle methods. J. Comput. Phys. 298, 280−304.

Ichikawa, H., Labrosse, S., 2010. Smooth particle approach for surface tension calculation in moving particle semi-implicit method. Fluid Dyn. Res. 42, 035503.

Ikejiri, S., Liu, J., Oka, Y., 2007. Simulation of a single bubble rising with hybrid particle mesh method. J. Nucl. Sci. Technol. 44 (6), 886−893.

Ishii, E., Ishikawa, T., Tanabe, Y., 2006. Spray simulation of fuel injector for automobile engines. Progr. Multiphase Flow Res. I 71−78 (in Japanese).

Shakibaeinia A., 2015. Mesh-free modeling of air-water interaction. E-proceedings of the 36th IAHR World Congress.

Shirakawa, N. 2000. Simulations of two-phase flows and jet flows with the particle interaction method, Ph.D. thesis, University of Tokyo.

Shirakawa, N., Horie, H., Yamamoto, Y., Tsunoyama, S., 2001a. Analysis of the void distribution in a circular tube with the two-fluid particle interaction method. J. Nucl. Sci. Technol. 38, 392−402.

Shirakawa, N., Horie, H., Yamamoto, Y., Okano, Y., Yamaguchi, A., 2001b. Analysis of jet flows with the two-fluid particle interaction method. J. Nucl. Sci. Technol. 38, 729−738.

Shirakawa, N., Yamamoto, Y., Horie, H., Tsunoyama, S., 2002. Analysis of flows around a BWR spacer by the two-fluid particle interaction method. J. Nucl. Sci. Technol. 39, 572−581.

Shirakawa, N., Yamamoto, Y., Horie, H., Tsunoyama, S., 2003. Analysis of subcooled boiling with the two-fluid particle interaction method. J. Nucl. Sci. Technol. 40, 125−135.

Wada, Y., Yuge, K., Nakamura, R., Tanaka, H., Nakamura, K., 2015. Dynamic analysis of ultrasonically levitated droplet with moving particle semi-implicit and distributed point source method. Jpn. J. Appl. Phys. 54, 07HE04.

Nakamura, K., 2016. 粒子法と分布点音源法による超音波不要液滴回転のシミュレーション. 超音波TECHNO 2016-11-12 (in Japanese).

Hattori, T., Hiai, D., Akaeke, S., Koshizuka, S., 2016. Improvement of wetting calculation model on polygon wall in the MPS method. Transact. JSME 82 (835), 15−00602 (in Japanese).

Ido, T., Inui, M., Tanaka, T., Tsuji, T., 2011. MPS simulation of wetting behavior of droplet. J. Soc. Powder Technol. Jpn 48, 822−828 (in Japanese).

Inagaki, K., 2015. Multi-physics particle method for the simulation of severe accidents in nuclear power plants. Transact. Atomic Energ. So. Jpn 14 (4), 249−260 (in Japanese).

Ishii, E., Sugii, T., 2012. Development of surface tension model for particle method. Transact. Jpn Soc. Mech. Eng. B 78, 1710−1725 (in Japanese).

Kondo, M., Koshizuka, S., Takimoto, M., 2007a. Surface tension model using inter-particle potential force in moving particle semi-implicit method. Transact. JSCES Paper No. 20070021 (in Japanese).

Kondo, M., Koshizuka, S., Suzuki, K., Takimoto M. 2007b. Surface tension model using inter-particle force in particle method. Proceedings of the FEDSM2007, 5th Joint ASME/JSME Fluids Engineering Conference, FEDSM2007-37192.

Michiwaki, Y., Kikuchi, T., Sonomura, M., 2014. Soy protein research. Japan 17, 32−36 (in Japanese).

Nodomi, S., 2015. Techno net 568, 16−21 (in Japanese) http://www.osaka-u.info/pdf/2015/04/top_article_3.pdf.

Saso, S., Mouri, M., 2017. Development of a simulation for the TIG arc welding phenomena. Technol. Rep. IHI 57 (1), (in Japanese).

Yoneda, M., Takimoto, M., 2007. Modeling and simulation for polymer electrolyte fuel cells. Technical Rep. Mizuho Inform. Res. Inst. 1 (1), (in Japanese).

Zhang, S., Morita, K., Shirakawa, N., Fkukuda, K., 2004. Simulation of the Rayleigh−Taylor instability with the MPS method,. Memoirs Faculty Eng. Kyushu Univ. 64.

CHAPTER 6

Advanced Techniques

Abstract

Advanced techniques for the Moving Particle Semi-implicit (MPS) method are explained. Liquid—solid phase change is represented by changing moving particles to fixed ones. Gas—liquid two-phase flow has been solved by a two-fluid model, MPS-MAFL, and a particle-mesh hybrid method. Turbulent flow is analyzed by a subparticle-scale turbulence model. Numerical techniques for suppressing pressure fluctuations are provided. High-order schemes are introduced. Parallel computing techniques for large-scale calculations are described: neighboring buckets, renumbering, and domain decomposition. Multiresolution techniques are explained. Concept of verification and validation (V&V) for modeling and simulation is briefly explained, and V&V examples for industrial applications are introduced: automobile, chemical engineering, metal engineering, and biomechanics.

Keywords: Liquid—solid phase change; gas—liquid two-phase flow; turbulence model; pressure fluctuation; higher-order scheme; parallel computing; multiresolution; verification and validation (V&V); industrial application

Contents

6.1	Liquid—Solid Phase Change Model	233
6.2	Gas—Liquid Two-Phase Flow and Phase Change Model	236
6.3	Turbulence	242
6.4	Suppression of Pressure Fluctuations	245
6.5	Higher-Order Schemes	247
6.6	Parallel Computing	250
6.7	Multiresolutions	254
6.8	V&V and Applications	259
	6.8.1 Verification and Validation	259
	6.8.2 Application to Automobile Industry	260
	6.8.3 Application to Chemical Engineering	265
	6.8.4 Application to Metal Engineering	268
	6.8.5 Application to Biomechanics	270
References		273

6.1 LIQUID—SOLID PHASE CHANGE MODEL

A melting/solidification model for the moving particle semiimplicit (MPS) method was developed by Ikeda (1999). More detailed

Moving Particle Semi-implicit Method
DOI: https://doi.org/10.1016/B978-0-12-812779-7.00006-0

formulation is described in Kawahara and Oka (2012). The governing equation with respect to enthalpy H

$$\frac{DH}{Dt} = k\nabla^2 T + Q \tag{6.1}$$

is solved for energy conservation, where k, T, and Q are thermal conductivity, temperature, and heat source, respectively. Temperature is given as a function of enthalpy

$$T = \begin{cases} T_m + \dfrac{H - H_{s0}}{\rho C_{ps}} & (H < H_{s0}) \\[2mm] T_m & (H_{s0} \le H \le H_{s1}), \\[2mm] T_m + \dfrac{H - H_{s1}}{\rho C_{pl}} & (H_{s1} < H) \end{cases} \tag{6.2}$$

where T_m is the melting temperature, H_{s0} and H_{s1} are the enthalpies at the start and the end of the melting, respectively, and C_{ps} and C_{pl} are the specific heats of the solid and liquid phases, respectively. Corresponding to Chen et al. (2014) a solid fraction γ is evaluated as

$$\gamma = \begin{cases} 1 & (H < H_{s0}) \\[2mm] \dfrac{H_{s1} - H}{H_{s1} - H_{s0}} & (H_{s0} \le H \le H_{s1}) \\[2mm] 0 & (H_{s1} < H) \end{cases} \tag{6.3}$$

Fig. 6.1 shows the temperature and the solid fraction with respect to the enthalpy in the liquid–solid phase change. When the enthalpy H is below H_{s0}, the phase is solid ($\gamma = 1$) and T increases with H. When the enthalpy H is between H_{s0} and H_{s1}, T keeps a constant value of T_m, where the solid fraction γ decreases from 1 to 0. When the enthalpy H is above H_{s1}, T increases with H as the liquid ($\gamma = 0$).

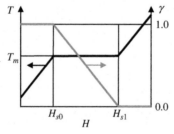

Figure 6.1 Liquid–solid phase change model: relation among enthalpy H, temperature T, and solid fraction γ.

In the MPS method, each particle has additional variables of H_i, T_i, and γ_i, where i is the particle number. In three variables, H_i is calculated by solving the transport equation of Eq. (6.1). The left-hand side of Eq. (6.1) is the Lagrangian derivative of H, which means that H_i is transported with moving particle i. The first term of the right-hand side of Eq. (6.1) is the Laplacian of T, which means the heat conduction. T_i is dependent on H_i as shown in Eq. (6.2). The Laplacian operator is discretized by the Laplacian model of the MPS method. γ_i is a dependent variable on H_i as shown in Eq. (6.3). When the solid fraction is above a certain value, the particle is fixed in space as a solid particle.

There are some variations of the melting/solidification model to incorporate the specific properties of the analyzed materials. For example the temperature is changed during the solidification. The temperature of solidification start and the temperature of solidification end are different. Viscosity increases with decrease of temperature or increase of the solid fraction. In this case the viscosity terms in the Navier—Stokes equations are implicitly discretized to keep the numerical stability.

The melting/solidification model has been applied to severe accident studies in the nuclear engineering (Kawahara and Oka, 2012; Chen and Oka, 2014; Chen et al., 2014; Li and Oka, 2014; Li et al., 2014; Chai et al., 2016; Li and Yamaji, 2017). In severe accidents of the nuclear reactors, we need to analyze the reactor core melting, relocation, spreading, and solidification by cooling. Fig. 6.2 illustrates an example (Chai et al., 2016). The molten core at a temperature of 2153 K was initially relocated

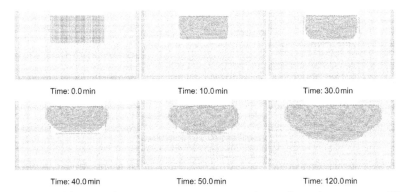

Time: 0.0 min   Time: 10.0 min   Time: 30.0 min

Time: 40.0 min   Time: 50.0 min   Time: 120.0 min

Figure 6.2 Ablation of concrete in severe accident of nuclear reactor (Chai et al., 2016).

to a cavity of concrete at a temperature of 293 K. The melting temperature of concrete was depending on the material composition. Since the molten core contained radioactive materials which produced heat for a long time, the surrounding concrete was ablated and the molten core pool expanded as shown in Fig. 6.2. This phenomenon is called Molten Core Concrete Interaction (MCCI). The shape of the pool was affected by the mixing model of the materials.

Saso et al. (2016) applied the melting/solidification model to welding analysis. Various phenomena were considered in the welding process: surface tension, Marangoni force, buoyancy, heat conduction, radiation heat transfer, and melting/solidification.

In the analysis of cell adhesion a model which was similar to the melting and solidification model was employed (Suzuki et al., 2006; Kamada et al., 2010). A fluid particle was attached to the wall when a certain condition was satisfied. An attached particle was released when another certain condition was satisfied. The shape of a microchannel changed due to the cell adhesion. Changing of a fluid particle to a solid particle is like solidification. This type of simulation was applied to thrombus after Fontan procedures by Sugimoto et al. (2015). Similar approach has been used for water—soil interaction problems in civil engineering.

6.2 GAS—LIQUID TWO-PHASE FLOW AND PHASE CHANGE MODEL

Ikeda et al. (1998) proposed a boiling model for the MPS method. Gas and liquid phases are analyzed by gas and liquid particles which have a common volume and different masses. Thus the mass of a gas particle is smaller than that of a liquid particle. The mass, momentum, and thermal energy conservation equations are solved for both phases:

$$\frac{D\alpha_g \rho_g}{Dt} = \Gamma_g, \tag{6.4}$$

$$\frac{D\alpha_l \rho_l}{Dt} = \Gamma_l, \tag{6.5}$$

$$\frac{D\alpha_g u_g}{Dt} = -\frac{1}{\rho_g}\nabla\alpha_g P + \nu_g \nabla^2 \alpha_g u_g + F_g + M_g, \tag{6.6}$$

$$\frac{D\alpha_l \mathbf{u}_l}{Dt} = -\frac{1}{\rho_l}\nabla\alpha_l P + \nu_l\nabla^2\alpha_l\mathbf{u}_l + \mathbf{F}_l + \mathbf{M}_l, \qquad (6.7)$$

$$\frac{D\alpha_g H_g}{Dt} = k_g\nabla^2\alpha_g T_g + Q_g, \qquad (6.8)$$

$$\frac{D\alpha_l H_l}{Dt} = k_l\nabla^2\alpha_l T_l + Q_l, \qquad (6.9)$$

where subscripts g and l denote gas and liquid, respectively. Phase fractions of gas and liquid are expressed by α_g and α_l, respectively. Γ, $\mathbf{F}$, $\mathbf{M}$, and Q are the mass source due to phase change, the external force such as gravity and surface tension, the momentum source term due to phase change, and the heat source, respectively. In Ikeda et al. (1998) the surface tension forces and the momentum source terms due to phase change are neglected.

In the phase change model the interfaces are locally assumed between the liquid and gas particles, as shown in Fig. 6.3. The temperature is assumed to be saturated on the interfaces. Heat transfer to interface i is evaluated from liquid particles q_{li} and gas particles q_{gi}. The heat energy is accumulated on the interfaces in each time step and stored in the liquid particle j possessing interfaces i. The mass transfer is obtained as the stored energy divided by the latent heat.

$$\frac{D\gamma_j}{Dt} h_{gl} = \sum_i (q_{gi} + q_{li}), \qquad (6.10)$$

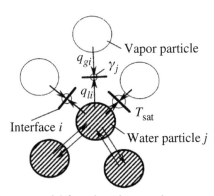

Figure 6.3 Phase change model from liquid to gas by generation of new gas particles on interfaces.

where γ_j and h_{gl} are the stored mass at particle j and the latent heat, respectively. This means that each liquid particle j has an additional variable γ_j. Once the stored mass exceeds that of one gas particle $\rho_g V$, where V is the particle volume, a new gas particle is generated on the interface and the mass of a gas particle is subtracted from the stored mass.

Boiling is modeled as generation of new gas particles. Since the volume expansion from a liquid to a gas is normally very large, the mass decrease of the liquid particles is neglected. For example, when saturated liquid water changes to saturated steam at the atmospheric pressure, the volume expands to approximately 1600 times. Therefore even if 16 steam particles are generated from one liquid particle, the mass decrease of the liquid particle is only 1%.

Particle simulation of two fluids that have a large density ratio is likely to be unstable on the interfaces. In Ikeda (1999) the pressure fields of the gas and the liquid were separately solved to keep the numerical stability. When the gas (light fluid) was solved, the liquid (heavy fluid) was treated as the fixed wall. When the liquid was solved, the interface with the gas was treated as the free surface where the gas pressure obtained in the previous time step was given as the Dirichlet boundary condition.

Impingement of jets of a hot molten metal on a cold water pool was analyzed by Ikeda et al. (1998). Boiling effect was considered by using the present boiling model. It was found that the jet penetration behavior was mainly governed by the density ratio between the jet and pool fluids.

Shirakawa et al. (2001a) used almost the same model for gas−liquid two-phase flow except for the surface tension. A potential force model, explained in Chapter 5, Surface Tension Models in Particle Methods, was employed for the surface tension.

Yoon et al. (1999a) developed another method to analyze single bubble behavior more accurately. Arbitrary Lagrangian−Eulerian (ALE) treatment, explained in Section 3.4, was investigated for the MPS method to keep the bubble interface smooth and stable. The surface tension is considered as the interparticle force between two interface particles. Only the liquid is discretized to liquid particles, while the gas is assumed uniform and no gas particles are introduced. The saturation temperature is assumed on the interface and heat conduction from the liquid is evaluated. Increase or decrease of the mass of the gas is obtained from the total heat conduction to the interface. This ALE method is called MPS-MAFL (meshless advection using flow-directional local-grid).

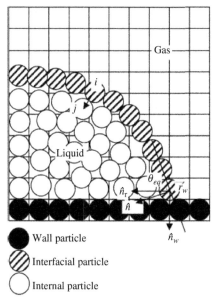

Figure 6.4 Surface tension and wettability of particle-mesh hybrid method.

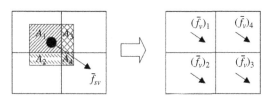

Figure 6.5 Transfer of surface tension force acting at a liquid particle to mesh.

Liu et al. (2005) developed a particle-mesh hybrid method for gas—liquid two-phase flow. The motivation was to analyze the liquid behavior more accurately and efficiently with considering the effect of gas which occupied the most of the space. The liquid phase is analyzed by the particles, whereas the gas phase is analyzed by the mesh. The main part of the development is the treatment of the surface tension and wettability (Fig. 6.4). The surface tension force acting at the liquid particles is distributed to the mesh (Fig. 6.5). Thus this method is like a particle-in-cell (PIC).

Another particle-mesh hybrid method was proposed by Ishii et al. (2006). The liquid phase was solved by both the particles and the mesh. The resultant new time values were interpolated between two solutions. The particles were necessary near the interfaces. The liquid phase far from the interfaces was solved by the mesh only. Collapse of a water column was solved by the present hybrid method. Liquid particles were located only near the free surface. Thus the complex motion involving the liquid fragmentation on the interface was successfully calculated though the main part of the liquid was efficiently solved by the mesh. The present hybrid method has been applied to a fuel injector of the automobiles.

Yoon et al. (1999b) analyzed single bubble rising in a stagnant fluid using MPS-MAFL. Coalescence of two bubbles was calculated using MPS-MAFL by Chen et al. (2011). Single bubble rising, bubble separation, and bubble coalescence were solved using the particle-mesh hybrid method by Ikejiri et al. (2007). Duan et al. (2017) analyzed the bubble behavior using both gas and liquid particles. Numerical instability was avoided by introducing continuous acceleration. The pressure gradient term is discretized to

$$
\left\langle \frac{1}{\rho} \nabla P \right\rangle_i = \frac{d}{n_0} \sum_{j \neq i} \left\{ \frac{2(P_j - P_i)(\mathbf{r}_j - \mathbf{r}_i)}{(\overline{\rho}_j + \overline{\rho}_i)|\mathbf{r}_j - \mathbf{r}_i|^2} \mathbf{C}_i w_{ij} \right\}
$$
$$
+ \frac{d}{n_0} \sum_{j \neq i} \left\{ \frac{2(P_i - P'_{i,\min})(\mathbf{r}_j - \mathbf{r}_i)}{\overline{\rho}_i |\mathbf{r}_j - \mathbf{r}_i|^2} \mathbf{C}_i w_{ij} \right\}, \tag{6.11}
$$

where $\mathbf{C}_i$ is a corrective matrix (Khayyer and Gotoh, 2011; Jeong et al., 2013)

$$
\mathbf{C}_i = \left[\frac{1}{n_0} \sum_{j \neq i} w_{ij} \left(\frac{\mathbf{r}_j - \mathbf{r}_i}{|\mathbf{r}_j - \mathbf{r}_i|} \otimes \frac{\mathbf{r}_j - \mathbf{r}_i}{|\mathbf{r}_j - \mathbf{r}_i|} \right) \right]^{-1}. \tag{6.12}
$$

In the second term of the right-hand side of Eq. (6.11), $P'_{i,\min}$ represents the minimum pressure of the same phase particles in the neighborhood of particle i. In Eq. (6.11) the density is included in the summation as the average of particles i and j. When particles i and j belong to different phases, averaging their densities makes the pressure gradient term continuous near the interface. In addition, density smoothing

$$
\overline{\rho}_i = \frac{1}{\displaystyle\sum_{j \neq i} w_{ij}} \sum_{j \neq i} \rho_j w_{ij}, \tag{6.13}
$$

is employed as well. These techniques successfully enhance numerical stability.

Bubble growth and departure on a heated wall in the nucleate boiling were analyzed by MPS-MAFL (Yoon et al., 2001; Heo et al., 2002; Chen et al., 2010). Heat and mass transfer on a bubble surface was calculated. Fig. 3.10 shows the calculation result of the nucleate boiling where the heated wall is 110°C and the bulk water is 96°C at the atmospheric pressure (Yoon et al., 2001). We can see that the initial small bubble rapidly grows in the hot boundary layer near the heated wall. The growth is decelerated by condensation in the bulk subcooled water when the bubble becomes larger. Finally, buoyancy force makes the bubble detached from the bottom wall. Heo et al. (2002) applied this boiling simulation to a Reactivity Initiated Accident (RIA) of the nuclear reactors. If the control rods are abruptly withdrawn from the nuclear reactor, the reactor power rapidly increases and transient boiling occurs on the fuel rod surface. Decrease of water density due to boiling is expected to reduce the power due to the negative feedback. The transient bubble growth was simulated with a heat flux of 2 MW/m^2 using MPS-MAFL to evaluate the time delay of the negative feedback. The simulation result agreed with the experiment. Duan et al. (2006) analyzed boiling as generation of new vapor particles using the boiling model of Ikeda et al. (1998).

Collapse of a bubble with condensation has been analyzed by MPA-MAFL (Tian et al., 2009, 2010a,b). No particles are located in the vapor phase. Heat transfer from the saturated interface to the subcooled water is calculated. Mass transfer from vapor to water is evaluated as the heat transfer divided by the latent heat.

Breakup of a moving liquid droplet in gas was analyzed by Nomura et al. (2001). The continuum surface force model for surface tension was developed and involved in the simulation. The breakup behavior is governed by the Weber number which represents the ratio of the disruptive hydrodynamic force over the stabilizing surface tension force. Breakup occurs when the Weber number is above the critical value (Nomura et al., 2001; Duan et al., 2003a,b). The details of the surface tension models are explained in Chapter 5, Surface Tension Models in Particle Methods.

Water droplet impingement on a water film, which takes place on the fuel pins in boiling water reactors, has been analyzed by the MPS method (Xie et al., 2004a,b, 2005). When the impingement speed is low, the droplet water is totally absorbed in the water film. On the other hand, when the impingement speed is high, new water droplets are generated and released from the water film by the impact of the droplet impinge-ment. The fuel pins should be covered with the water film to cool them. Thus it is important to estimate the droplet water is added to the liquid film or not when the droplet impinges on the water film.

Liquid droplet impingement (LDI) occurs around bends and orifices of pipes in which high-speed steam flow is accompanied by liquid droplets. Repeated LDI may damage the pipe inner surface. The pressure load due to LDI has been analyzed by the MPS method (Arai and Koshizuka, 2009; Xiong et al., 2010, 2011, 2012). Pressure propagation inside the water droplet is calculated by the compressible−incompressible unified algorithm described in Section 3.1. Figs. 3.1 and 3.2 show the pressure distributions inside the droplets (Xiong et al., 2010, 2011) when the pipe inner surface is dry and covered with a thin water film, respec-tively. We can observe pressure waves propagating inside the droplets.

Particle-mesh hybrid methods have also been used for the gas−liquid two-phase flows. Droplet entrainment from the liquid film to annular gas flow was analyzed by Yun et al. (2010). Droplet deposition onto a liquid film was analyzed by Erkan et al. (2015).

Droplet behavior in a microchannel was studied by Harada et al. (2005). Binary droplet collision was analyzed by Sun et al. (2009). Droplet behavior on a wall was analyzed by Hattori et al. (2016) to improve the accuracy of the potential model of surface tension and wetta-bility combined with the polygon wall boundary condition. Ishii et al. (2011) and (2014) investigated the secondary droplet breakup by using a particle-mesh hybrid method. Jet breakup was also analyzed by the MPS method (Koshizuka et al., 1999; Ikeda et al., 2001; Shirakawa et al., 2001b; Shibata et al., 2004). The droplet size distribution after the jet breakup was obtained (Shibata et al., 2004). Gas−liquid two-phase flows were simulated using both liquid and gas particles by Horie et al. (2001) and Shirakawa et al. (2003). Various multiphase thermal-hydraulic problems for nuclear safety were solved and compared with experiments for the code development (Morita et al., 2011).

6.3 TURBULENCE

A subparticle-scale (SPS) turbulence model was first introduced into the MPS method by Gotoh et al. (2000, 2001) for Large Eddy Simulation (LES). SPS was similarly named from the subgrid-scale (SGS) turbulence model in the grid methods.

The velocity vector u_α is decomposed into a spatially averaged value $\bar{u}_\alpha$ and a fluctuating value u'_α as

$$u_\alpha = \bar{u}_\alpha + u'_\alpha, \tag{6.14}$$

where subscript α represents the spatial component changing from 1 to 3. All the variables, the velocity vector and the pressure, in the Navier–Stokes equations are replaced by spatially averaged values and fluctuating values, and the obtained Navier–Stokes equations are spatially averaged. Then, we have

$$\frac{D\bar{u}_\alpha}{Dt} = -\frac{1}{\rho}\frac{\partial \bar{P}}{\partial x_\alpha} + \nu \frac{\partial}{\partial x_\beta}\left(\frac{\partial \bar{u}_\alpha}{\partial x_\beta} + \frac{\partial \bar{u}_\beta}{\partial x_\alpha}\right) - \frac{\partial(\overline{u'_\alpha u'_\beta})}{\partial x_\beta}. \tag{6.15}$$

A new term appears in the right-hand side of Eq. (6.15), where $\overline{u'_\alpha u'_\beta}$ is called the Reynolds stress. It is derived from the convection term which is nonlinear and included in the Lagrangian derivative.

The Reynolds stress is modeled as

$$-\overline{u'_\alpha u'_\beta} = \nu_t\left(\frac{\partial \bar{u}_\alpha}{\partial x_\beta} + \frac{\partial \bar{u}_\beta}{\partial x_\alpha}\right) - \frac{2}{3}k\delta_{\alpha\beta}, \tag{6.16}$$

where k is turbulence kinetic energy

$$k = \frac{1}{2}\overline{u'_\alpha u'_\alpha}, \tag{6.17}$$

and δ is the delta function

$$\delta_{\alpha\beta} = \begin{cases} 1 & \alpha = \beta \\ 0 & \alpha \neq \beta \end{cases}. \tag{6.18}$$

In Eq. (6.16), ν_t is eddy viscosity and modeled as

$$\nu_t = (C_s\Delta)^2\left\{\left(\frac{\partial \bar{u}_\alpha}{\partial x_\beta} + \frac{\partial \bar{u}_\beta}{\partial x_\alpha}\right)\frac{\partial \bar{u}_\alpha}{\partial x_\beta}\right\}^{1/2}, \tag{6.19}$$

where C_s and Δ are the Smagorinsky constant and the particle size, respectively. In the MPS method, Δ can be the particle spacing l_0. Combining Eqs. (6.15) and (6.16), we have

$$\frac{D\bar{u}_\alpha}{Dt} = -\frac{1}{\rho}\frac{\partial \bar{P}_t}{\partial x_\alpha} + (\nu + \nu_t)\frac{\partial}{\partial x_\beta}\left(\frac{\partial \bar{u}_\alpha}{\partial x_\beta} + \frac{\partial \bar{u}_\beta}{\partial x_\alpha}\right), \tag{6.20}$$

$$\overline{P}_t = \overline{P} + \frac{2}{3}\rho k \qquad (6.21)$$

In Eq. (6.20) the viscosity is increased from ν to $\nu + \nu_t$. Roughly speaking the viscosity is larger in turbulent flow by ν_t though ν_t is a variable evaluated by Eq. (6.19).

This SPS turbulence model has been used in the MPS method for turbulent flows (Arai et al., 2013; Duan and Chen, 2015; Li et al., 2016). In Arai et al. (2013) a wall function is used as the wall boundary condition of turbulence. The fluid particles which are close to the wall are forced to be on the universal logarithmic law called wall function. Friction force to the wall is evaluated using the wall function. Roughly speaking, this wall boundary condition is between the nonslip and the free-slip. When the wall function is used, the spatial resolution is not necessary to be finer near the wall since the wall function is universally assumed in the boundary layer between the fluid particle and the wall. Thus it is fitted to the particle method where the spatial resolution is usually uniform and isotropic. In many of complex three-dimensional problems in industry, it is very difficult to make the spatial resolution finer near the wall. The wall function is useful for these problems.

Arai et al. (2013) investigated the fluctuating motion originated from the turbulence and the Lagrangian description. It has been known that the particle motion is disturbed by fluctuation derived from the Lagrangian description of the particle method. Particularly the pressure field is likely to involve spatial and temporal fluctuations. This undesirable characteristic is discussed in the next section. Channel flows were calculated with Reynolds numbers (Re) of 0.5 and 50,000 (Arai et al., 2013). The fluctuating motion was only derived from the Lagrangian description in Re = 0.5, while it was derived from both the Lagrangian description and turbulence in Re = 50,000. The fluctuating motion was substantially larger in Re = 50,000 than that in Re = 0.5 in most of the domain. This means that turbulence can be analyzed by the MPS method. However, both fluctuations in Re = 50,000 and Re = 0.5 were not largely different near the wall. This might be due to the wall boundary condition in which wall particles were arranged and the bumpy wall was presumed. We can say that fluctuating motion near the wall is affected by the wall boundary condition using the bumpy wall particles and turbulent motion is degraded.

6.4 SUPPRESSION OF PRESSURE FLUCTUATIONS

The MPS method suffers violent numerical pressure fluctuations. They are both spatial and temporal. These fluctuations do not observed in the finite volume method (FVM). Let us compare the pressure Poisson equations in the MPS method and FVM:

$$\text{MPS:} \quad \left\langle \nabla^2 P^{k+1} \right\rangle_i = -\frac{\rho}{\Delta t^2} \frac{n_i^* - n^0}{n^0}, \tag{6.22}$$

$$\text{FVM:} \quad \left\langle \nabla^2 P' \right\rangle_i = \frac{\rho}{\Delta t} \nabla \cdot \boldsymbol{u}_i^*, \tag{6.23}$$

where subscript i denotes the particle number in MPS and the node number in FVM. The right-hand sides are largely different between Eqs. (6.22) and (6.23). The deviation of the temporary particle number density n_i^* from the constant value n^0 is used in MPS, while the divergence of the temporary velocity is used in FVM. This difference may be the main reason of the numerical fluctuations of pressure in MPS.

We can see a small difference in the left-hand sides as well. P^{k+1} is used in MPS, while P' is used in FVM. P' is the modification of pressure

$$P^{k+1} = P^k + P'. \tag{6.24}$$

In FVM the pressure gradient term is explicitly calculated using P^k for the temporary velocity $\boldsymbol{u}^*$. Thus the pressure Poisson equation is made for P' and the pressure gradient term using P' is calculated for the new time velocity $\boldsymbol{u}^{k+1}$. In MPS the pressure gradient term is not calculated for the temporary velocity $\boldsymbol{u}^*$. The pressure Poisson equation is made for P^{k+1} and the pressure gradient term using P^{k+1} is calculated for the new time velocity $\boldsymbol{u}^{k+1}$. It is know that the calculation is less stable when the pressure gradient term is calculated using P^k for the temporary velocity in MPS.

Using the particle number density as the source term of the pressure Poisson equation is a specific point of MPS. Being the fluid density constant is direct expression of incompressibility and preferable to avoid the accumulation of the error of incompressibility. For instance, if there is a deviation from n^0, the source term of the pressure Poisson equation is not zero and the pressure field is given to restore the particle number density to n^0. If the divergence of the temporary velocity is used in a particle method, the particle number density will not be kept at n^0. On the other hand the particle number density would be sensitive to the configuration of the moving particles. The particle number density is likely to change due to the motion of particles. This makes significant fluctuations of pressure. Besides the right-hand side of Eq. (6.22) involves $1/(\Delta t)^2$, which leads to larger fluctuations as Δt decreases.

Hibi and Yabushita (2004) and Sueyoshi (2005) proposed modified algorithms where the pressure Poisson equation was solved twice by different source terms using the particle number density and the divergence of the velocity. The pressure Poisson equation using the particle number density is used for the particle movement. This is good to avoid the accumulation of errors of incompressibility as the usual MPS method. Whereas the pressure field is given as the solution of the pressure Poisson equation using the divergence of the velocity. This gives us a smoother pressure field. This type of algorithm was also deduced by Suzuki et al. (2007) as a symplectic scheme (Section 3.3). However, the calculation time is doubled to solve the pressure Poisson equation twice.

Kondo and Koshizuka (2008) proposed a modified source term of the pressure Poisson equation

$$\left\langle \nabla^2 P^{k+1} \right\rangle_i = -\frac{\rho}{\Delta t^2} \left[\alpha \frac{n_i^* - 2n_i^k + n_i^{k-1}}{n^0} + \beta \frac{n_i^k - n_i^{k-1}}{n^0} + \gamma \frac{n_i^k - n^0}{n^0} \right].$$

$$(6.25)$$

If $\alpha = \beta = \gamma = 1.0$, Eq. (6.25) will be the same as the usual pressure Poisson equation of MPS (Eq. (6.22)). The pressure field is smoother when proper parameters are chosen. Tanaka and Masunaga (2010) proposed another source term,

$$\left\langle \nabla^2 P^{k+1} \right\rangle_i = \frac{\rho}{\Delta t} \nabla \cdot \boldsymbol{u}_i^* - \gamma \frac{\rho}{\Delta t^2} \frac{n_i^k - n^0}{n^0}.$$

$$(6.26)$$

This is a more straightforward expression of mixture of two types of the pressure Poisson equations Eqs. (6.22) and (6.23). Compromising of the pressure fluctuations and the error accumulation is controlled by parameter γ. Kondo and Koshizuka (2011) showed that change of the particle number density from time k to $*$ was equivalent to the divergence of the velocity

$$\frac{n_i^* - n_i^k}{n^0 \Delta t} = -\nabla \cdot \boldsymbol{u}_i^*. \tag{6.27}$$

Thus we can say that the methods of Kondo and Koshizuka (2008) and Tanaka and Masunaga (2010) are almost the same. This type of the pressure Poisson equation was used in Khayyer and Gotoh (2011) and Lee et al. (2011).

In the usual MPS method, decrease of the particle number density is used to detect the free surface

$$n_i^* < \beta n^0, \tag{6.28}$$

where parameter $\beta < 1.0$ is introduced. Dirichlet boundary condition $P_i^{k+1} = 0$ is given where Eq. (6.28) is satisfied. However, this condition is often and unexpectedly satisfied inside the fluid and $P_i^{k+1} = 0$ appears. This is also a reason of the pressure fluctuations. Gotoh et al. (2009) proposed another sophisticated condition for the free surface to avoid the unexpected detection inside the fluid. This new condition was effective to suppress the pressure fluctuations. Improved free-surface boundary conditions have been employed in other studies (Tanaka and Masunaga, 2010; Shibata et al., 2015; Tsukamoto et al., 2016).

Higher-order schemes are also effective in suppressing the pressure fluctuations. The corrective matrix (Suzuki, 2007) was applied to improve the numerical stability by Khayyer and Gotoh (2011). The higher-order schemes are explained more in the next section. Recent studies for suppressing the pressure fluctuations were reviewed in Tsukamoto et al. (2016) and Gotoh and Khayyer (2016).

6.5 HIGHER-ORDER SCHEMES

In the finite difference method (FDM), higher-order schemes can be constructed by using a Taylor expansion. Distributions of quantities are represented by a polynomial involving unknown coefficients. Here, a Taylor expansion of quantity ϕ_j to the first order is provided around particle i as

$$\phi_j = \phi_i + \nabla\phi|_i \cdot (\boldsymbol{r}_j - \boldsymbol{r}_i) + O(|\boldsymbol{r}_j - \boldsymbol{r}_i|^2). \tag{6.29}$$

In Eq. (6.29), we need to care that $\nabla\phi|_i$ is a vector of unknown coefficients representing the gradient of ϕ at particle i. The number of the unknown coefficients is that of space dimensions. Omitting the error term, we can obtain the following relationship from Eq. (6.29)

$$\frac{\phi_j - \phi_i}{|\boldsymbol{r}_j - \boldsymbol{r}_i|^2}(\boldsymbol{r}_j - \boldsymbol{r}_i) = \left(\nabla\phi\Big|_i \cdot \frac{\boldsymbol{r}_j - \boldsymbol{r}_i}{|\boldsymbol{r}_j - \boldsymbol{r}_i|}\right)\frac{\boldsymbol{r}_j - \boldsymbol{r}_i}{|\boldsymbol{r}_j - \boldsymbol{r}_i|} = \left(\frac{\boldsymbol{r}_j - \boldsymbol{r}_i}{|\boldsymbol{r}_j - \boldsymbol{r}_i|} \otimes \frac{\boldsymbol{r}_j - \boldsymbol{r}_i}{|\boldsymbol{r}_j - \boldsymbol{r}_i|}\right) \cdot \nabla\phi\Big|_i.$$

(6.30)

Furthermore,

$$\frac{1}{n^0}\sum_{j\neq i}\frac{\phi_j - \phi_i}{|\boldsymbol{r}_j - \boldsymbol{r}_i|^2}(\boldsymbol{r}_j - \boldsymbol{r}_i)w(|\boldsymbol{r}_j - \boldsymbol{r}_i|)$$

$$= \frac{1}{n^0}\sum_{j\neq i}w(|\boldsymbol{r}_j - \boldsymbol{r}_i|)\left(\frac{\boldsymbol{r}_j - \boldsymbol{r}_i}{|\boldsymbol{r}_j - \boldsymbol{r}_i|} \otimes \frac{\boldsymbol{r}_j - \boldsymbol{r}_i}{|\boldsymbol{r}_j - \boldsymbol{r}_i|}\right) \cdot \nabla\phi\Big|_i.$$

(6.31)

Then the unknown coefficients are given by

$$\nabla\phi\Big|_i = \left[\frac{1}{n^0}\sum_{j\neq i}w(|\boldsymbol{r}_j - \boldsymbol{r}_i|)\left(\frac{\boldsymbol{r}_j - \boldsymbol{r}_i}{|\boldsymbol{r}_j - \boldsymbol{r}_i|} \otimes \frac{\boldsymbol{r}_j - \boldsymbol{r}_i}{|\boldsymbol{r}_j - \boldsymbol{r}_i|}\right)\right]^{-1}$$

$$\cdot \left[\frac{1}{n^0}\sum_{j\neq i}\frac{\phi_j - \phi_i}{|\boldsymbol{r}_j - \boldsymbol{r}_i|^2}(\boldsymbol{r}_j - \boldsymbol{r}_i)w(|\boldsymbol{r}_j - \boldsymbol{r}_i|)\right]$$

(6.32)

Another expression is

$$\nabla\phi\Big|_i = \left[\frac{1}{n^0}\sum_{j\neq i}\frac{\phi_j - \phi_i}{|\boldsymbol{r}_j - \boldsymbol{r}_i|^2}\boldsymbol{C}_i(\boldsymbol{r}_j - \boldsymbol{r}_i)w(|\boldsymbol{r}_j - \boldsymbol{r}_i|)\right],$$

(6.33)

where

$$\boldsymbol{C}_i = \left[\frac{1}{n^0}\sum_{j\neq i}w(|\boldsymbol{r}_j - \boldsymbol{r}_i|)\left(\frac{\boldsymbol{r}_j - \boldsymbol{r}_i}{|\boldsymbol{r}_j - \boldsymbol{r}_i|} \otimes \frac{\boldsymbol{r}_j - \boldsymbol{r}_i}{|\boldsymbol{r}_j - \boldsymbol{r}_i|}\right)\right]^{-1},$$

(6.34)

is called a corrective matrix (Khayyer and Gotoh, 2011).

The right-hand side of Eq. (6.32) has two parts: the former part is the inverse of a tensor and the latter part is the gradient model of MPS. If the tensor in the former part is the unit tensor, Eq. (6.32) will be the same as the gradient model of MPS. This appears where the particles are arranged on a square grid in two dimensions or on a cubic grid in three dimensions. This shows that the spatial accuracy of the gradient model of MPS is lower than the first order in a general arrangement of the particles.

Applying Eq. (6.32) makes the MPS method to be first order in an arbitrary arrangement of the particles.

In the FDM the order of accuracy is assured in the square or cubic grid. Higher-order schemes are made for the uniform grid. When the grid is not uniform or deformed, a coordinate transformation technique is added and the higher-order schemes are applied to the transformed space. In this case the higher-order accuracy is not guaranteed in the physical space. This situation is similar to the original MPS method without Eq. (6.32).

Eq. (6.32) can be deduced from the method of least squares as well. The weighted squared residual from the first-order Taylor expansion is

$$S = \sum_{j \neq i} [\nabla\phi|_i \cdot (r_j - r_i) - (\phi_j - \phi_i)]^2 \frac{w(|r_j - r_i|)}{n^0}. \tag{6.35}$$

Derivatives of S with respect to the unknown coefficients are zero

$$\frac{\partial S}{\partial \nabla\phi|_i} = \frac{\partial}{\partial \nabla\phi|_i} \sum_{j \neq i} \left[\nabla\phi|_i \cdot (r_j - r_i) - (\phi_j - \phi_i)\right]^2 \frac{w(|r_j - r_i|)}{n^0} = 0. \tag{6.36}$$

This equation is reduced to

$$\sum_{j \neq i} \left[\nabla\phi|_i \cdot (r_j - r_i) - (\phi_j - \phi_i)\right] \frac{w(|r_j - r_i|)}{n^0} (r_j - r_i) = 0. \tag{6.37}$$

Thus Eq. (6.32) is obtained. This means that Eq. (6.32) is the first-order approximation of the method of least squares.

Eq. (6.32) was derived by Suzuki (2007). Iribe and Nakaza (2010) and Khayyer and Gotoh (2011) used this formulation. The corrective matrix was employed in Duan et al. (2017) for gas–liquid two-phase flow analysis as shown in Eq. (6.11).

Higher-order schemes have been formulated by using the Taylor expansion and the method of least squares (Tamai et al., 2013). The number of unknown coefficients are determined when the order of the Taylor expansion is given. The unknown coefficients can be evaluated if the number of particles is larger than that of the unknown coefficients. Thus an arbitrary order scheme can be formulated in an arbitrary arrangement of particles. In Tamai and Koshizuka (2014) a scaling parameter was introduced to keep the condition number low. In addition, a new formulation corresponding to the compact scheme (Lele, 1992) was proposed. Fig. 6.6 shows the

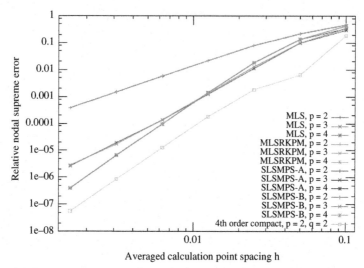

Figure 6.6 Error convergence of standard LSMPS schemes and 4th-order LSMPS compact scheme (Tamai and Koshizuka, 2014).

convergence of the error with respect to spatial resolution. The orders of the convergence agree with the orders of the proposed schemes. The compact scheme shows a lower error. These generalized schemes are called least squares MPS (LSMPS).

Hu et al. (2017) applied LSMPS to the convection term in the ALE description. ALE is explained in Section 3.4 as one of the extended algorithms. In the convection term, first-order spatial differentiations are necessary to discretize. Hu et al. (2017) added an upwind concept into LSMPS. Steady flow in a square cavity was analyzed by the upwind least square interpolation (ULSI) with the fully Eulerian description; all particles are fixed in space (Fig. 6.7). The obtained flow velocity distributions along the center lines are depicted in Fig. 6.8 with a Reynolds number of 3200. We can see good agreement between the solution of ULSI and that of Ghia et al. (1982).

6.6 PARALLEL COMPUTING

In the straightforward implementation of the MPS method the most time-consuming process is to search the neighboring particles. A particle interacts with its neighboring particles which are searched by evaluating the distance between two particles. When the distance is less than the

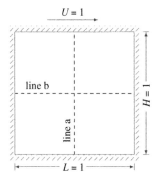

Figure 6.7 Square cavity problem; top wall moving toward right and left, right and bottom walls fixed (Hu et al., 2017).

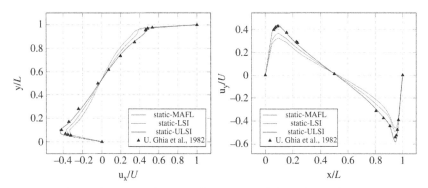

Figure 6.8 Flow velocity distributions of u on line a (left) and v on line b (right) with Re = 3200 and 200 × 200 particles (Hu et al., 2017).

effective radius r_e as shown in Fig. 1.2, they are the neighbors each other. The calculation of the distances between all the pairs of particles requires the order of $N^{2.0}$ operations, where N is the number of particles. On the other hand the explicit calculations of the particle interaction models in the MPS method are $N^{1.0}$. A typical iteration method to solve the simultaneous linear equations approximately requires $N^{1.5}$. Only the neighboring search needs the scaling of $N^{2.0}$. Though the calculation of the distances is simple, the scaling of $N^{2.0}$ makes the calculation time dominant for a larger N.

The search of the neighboring particles is an additional process in the particle method and it is not necessary in the grid methods because the neighbors are explicitly illustrated by the grid. Practically the method which requires the order of $N^{2.0}$ cannot be used for large-scale problems.

Gotoh et al. (2003) introduced buckets where the particles were registered in space as shown in Fig. 6.9. The space is divided to small buckets; usually, the buckets are cubes of the same size. A particle is involved in a bucket. The calculation of correspondence between the particles and the buckets is the order of $N^{1.0}$. The neighboring search can be limited to the particles in the neighboring buckets within the effective radius r_e. This search calculation has the order of $N^{1.0}$. Thus the calculation of the neighboring search is reduced from $N^{2.0}$ to $N^{1.0}$. This scaling is smaller than that of the solver. This technique is indispensable for large-scale calculations of the particle methods.

For parallel computing of the MPS method, Iribe et al. (2006) solved the simultaneous linear equations of the pressure Poisson equation using the conjugate gradient method with multiple processors. The particles were divided to multiple processors based on their numbers. In this case the particles in one processor are initially localized, and communication with the other processors is not large. As the particles move, they spread in space and the interaction with many neighboring particles handled by the other processors is necessary to calculate. This increases communication among the processors, and efficiency of the parallel computing is deteriorated. The locality of the particles in each processor is restored by introducing renumbering of the particles (Iribe et al., 2008, 2010). The numbers of particles are ordered again, for example, from the left to the right and from the bottom to the top.

Murotani et al. (2012) employed domain decomposition for parallel computing. The domain decomposition technique has been widely used in the grid methods and it was applied to the buckets in the MPS method (Fig. 6.10). In this case the communication among the processors is not

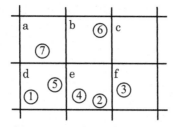

Figure 6.9 Buckets for neighboring search; (a–f) buckets, (1–7) particles.

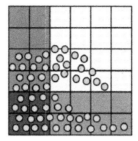

Figure 6.10 Domain decomposition technique based on buckets for parallel computing in the particle methods.

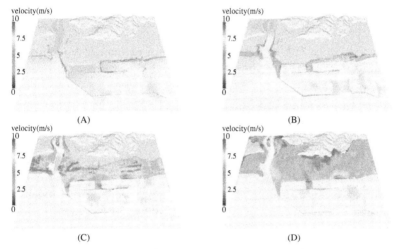

Figure 6.11 Tsunami run-up analysis on a coastal city using the wave of Great East Japan Earthquake in 2011 (color represents flow velocity) (Murotani et al., 2014). (A) 200 s; (B) 300 s; (C) 400 s; (D)600 s.

increased so much because the domain and its neighboring buckets are fixed in space. However, the number of particles in a domain changes as the particles move. Thus domain decomposition is performed again during the calculation when the load balance among the processors is deteriorated.

Fig. 6.11 is an example of a large-scale parallel computing using the domain decomposition technique (Murotani et al., 2014). It is a three-dimensional tsunami run-up analysis on a coastal city. The tsunami wave is assumed as that of the Great East Japan Earthquake in 2011.

In this example, the explicit algorithm using pseudo-compressibility explained in Section 3.2 is used. The number of particles is 260 million. The physical time simulated in this example is 800 s. The computation time is about 7 days using 96 processors of FX10 in the University of Tokyo. We can see that almost the whole city except for the hills is flooded at 600 s. Particularly, seawater is accumulated at the mouth of the river and seawater runs up along the river.

Special techniques of parallel computing on graphics processing units have been studied for the MPS method (Harada et al., 2007a,b,c, 2008a, b,c; Gotoh et al., 2010).

6.7 MULTIRESOLUTIONS

Multiresolution techniques are very effective in reducing the computation time of MPS simulations. Multiresolution techniques reduce the required number of particles by using nonuniform spatial resolutions, and allow us to reduce the computation time. This is because the calculation amount decreases as the number of particles decreases.

There are various multiresolution techniques for the MPS method. For example, Ikeda (1999) and Yoon et al. (1999a,b) developed partial derivative models of the MPS method for nonuniform particle-size simulations. There are also some variations of their nonuniform models. We would like to explain one of their models below. Each particle i has its own length l_i and volume. The volume of the i-th particle V_i is defined as follows:

$$V_i = (l_i)^d, \qquad (6.38)$$

where d is the number of spatial dimensions. The particle number density, gradient, and Laplacian are expressed with the volume of each particle as follows:

$$n_i = \sum_{j \neq i} \frac{V_j w(|\boldsymbol{r}_j - \boldsymbol{r}_i|, r_{ei}) + V_i w(|\boldsymbol{r}_j - \boldsymbol{r}_i|, r_{ej})}{2V_i}, \qquad (6.39)$$

$$\langle \nabla \phi \rangle_i = \frac{d}{V_i n^0} \sum_{j \neq i} \left[\frac{\phi_j - \phi_i}{|\boldsymbol{r}_j - \boldsymbol{r}_i|^2} (\boldsymbol{r}_j - \boldsymbol{r}_i) V_j w(|\boldsymbol{r}_j - \boldsymbol{r}_i|, r_{ei}) \right], \qquad (6.40)$$

$$\langle \nabla^2 \phi \rangle_i = \frac{2d}{V_i \Lambda_i} \sum_{j \neq i} \left[(\phi_j - \phi_i) \frac{V_j w(|\mathbf{r}_j - \mathbf{r}_i|, r_{ei}) + V_i w(|\mathbf{r}_j - \mathbf{r}_i|, r_{ej})}{2} \right],$$

(6.41)

where r_{ei} and r_{ej} are the radii of the effective domains for particle i and j, respectively. The scalar parameter Λ_i in Eq. (6.41) is calculated as follows:

$$\Lambda_i = \sum_{j \neq i} \left[|\mathbf{r}_j - \mathbf{r}_i|^2 \frac{V_j w(|\mathbf{r}_j - \mathbf{r}|, r_{ei}) + V_i w(|\mathbf{r}_j - \mathbf{r}|, r_{ej})}{2V_i} \right].$$

(6.42)

Yoon et al. (2001) applied a nonuniform particle model to the ALE method and solved a rising bubble as shown in Fig. 6.12. We find that the spatial resolution near the bubble is high because particles are densely distributed around the bubble.

Shibata et al. (2016) developed an ellipsoidal particle model for particle methods. Ellipsoidal particles can enhance the spatial resolution in given directions. In their model the governing equations are transformed

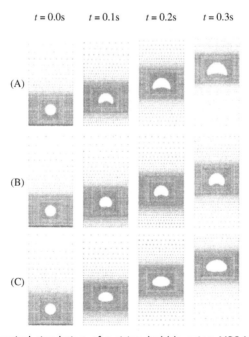

Figure 6.12 Numerical simulation of a rising bubble using MPS-MALE (Yoon et al., 2001). (A) Spherical cap; (B) Dimple; (C) Ellipsoid.

so that we can deal with ellipsoidal particles as if they are spherical particles. The Laplacian model of the MPS method was extended for ellipsoidal particles as follows:

$$
\langle \nabla^2 \phi \rangle_i^k = \frac{2d}{\hat{n}^0} \sum_{j \neq i} \left\{ \frac{1}{\lambda_x^0} \frac{(x_j^k - x_i^k)^2}{|\mathbf{r}_j^k - \mathbf{r}_i^k|^2} + \frac{1}{\lambda_y^0} \frac{(y_j^k - y_i^k)^2}{|\mathbf{r}_j^k - \mathbf{r}_i^k|^2} \right.
$$
$$
\left. + \frac{1}{\lambda_z^0} \frac{(z_j^k - z_i^k)^2}{|\mathbf{r}_j^k - \mathbf{r}_i^k|^2} \right\} (\phi_j^k - \phi_i^k) w(|\hat{\mathbf{r}}_j^k - \hat{\mathbf{r}}_i^k|),
$$

(6.43)

where $\hat{\mathbf{r}}_i^k$ is the transformed coordinate of the i-th particle at time step k. The parameters λ_x^0, λ_y^0, and λ_z^0 are scalar parameters for calibrating the calculation results of the Laplacian, and are determined by solving simultaneous equations at the initial setting. A droplet deformation was simulated by the ellipsoidal particle model. The simulated fluid shapes are shown in Fig. 6.13. At the initial state, particles were arranged at even

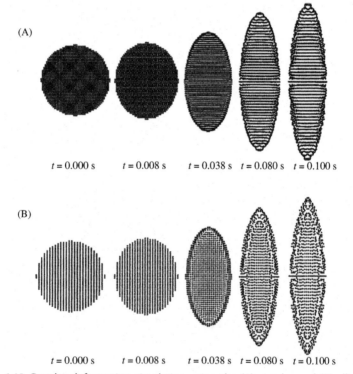

(A)

$t = 0.000$ s $t = 0.008$ s $t = 0.038$ s $t = 0.080$ s $t = 0.100$ s

(B)

$t = 0.000$ s $t = 0.008$ s $t = 0.038$ s $t = 0.080$ s $t = 0.100$ s

Figure 6.13 Droplet deformation simulation using the (A) circular and (B) ellipsoidal particle models (Shibata et al., 2016). (A) $s_x{:}s_y = 1{:}1$; (B) $s_x{:}s_y = 2{:}1$.

intervals, in x and y directions. In the case where the aspect ratio $s_x:s_y$ was 2:1, the initial spacing between particles in the x-direction was twice as long as the y-direction. That is, particles were elliptical, longer in the x-direction. After starting the simulation the droplet deformed because of the velocity distribution given at the initial state. We find that the simulation results of ellipsoidal particles (Fig. 6.13B) were almost the same as those of circular particles, whose aspect ratio is $s_x:s_y = 1:1$ (Fig. 6.13A).

Shibata et al. (2017) developed the overlapping particle technique (OPT), which is a multispatial resolution method for particle methods. The conceptual image is shown in Fig. 6.14. In the OPT, the simulation domain is divided into subdomains with their own spatial resolution. The subdomains are connected by mapping the velocities and pressure fields. On the boundaries between subdomains, inlet—outlet boundaries are located to generate particles or delete particles in accordance with the flow direction. Fig. 6.15A shows a dam-break simulation by the OPT. The left subdomain was simulated at low spatial resolution, whereas the right subdomain was simulated at high spatial resolution. We find that the simulation result of the OPT was almost the same as the that of fine particle simulation without OPT (Fig. 6.15B).

There are other studies about multiresolution simulation of the MPS method. For example, Tang et al. (2016) simulated 3D violent free-surface flows by a multiresolution MPS method. Tanaka et al. (2009) and Chen et al. (2016) developed multiresolution methods for the MPS

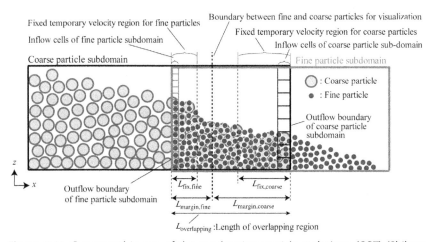

Figure 6.14 Conceptual image of the overlapping particle technique (OPT) (Shibata et al., 2017).

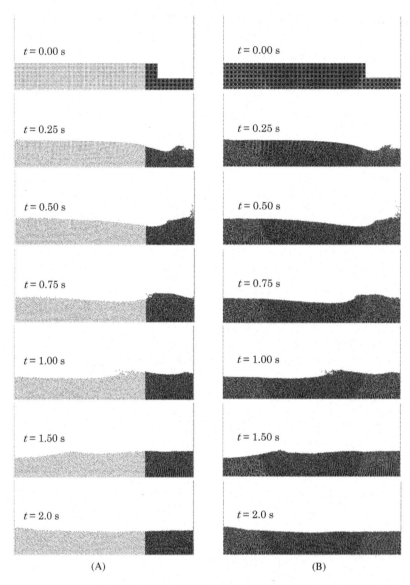

Figure 6.15 Dam-break simulation using the overlapping particle technique (Shibata et al., 2017). (A) With OPT; (B) without OPT

method on the basis of algorithms for dynamic particle coalescing and splitting. Jeong et al. (2011) developed a multiscale model which uses an anisotropic weighting function for the Laplacian model of the MPS method.

6.8 V&V AND APPLICATIONS

6.8.1 Verification and Validation

Verification and validation (V&V) for modeling and simulation (M&S) is a methodology to enhance credibility of engineering simulation. V&V for M&S is a general methodology and not limited to the particle methods. The basic idea of V&V for M&S is illustrated in Fig. 6.16. When we are requested to solve an engineering simulation in the real world, a conceptual model is first constructed. In this conceptual model, necessary laws of physics are chosen for the intended use. For instance, Navier–Stokes equations are chosen for fluid dynamics or Hooke's law is chosen for elastic solid dynamics. Unnecessary laws are excluded.

The conceptual model is usually represented by differential equations. They are discretized and programmed to be calculated in a computer. We also need boundary and initial conditions and calculation parameters. The computer program with input data involving the boundary conditions and the calculation parameters is called computational model.

The calculation results are obtained by using the computational model. On the other hand, analytical solutions can be deduced from the differential equation in simple geometries. Quantitative comparison between the analytical solutions and the calculation results is called verification. In the verification, errors derived from the spatial and temporal discretization are assessed. Bugs in the program are also checked. When the errors are within the tolerance, the verification is successfully carried

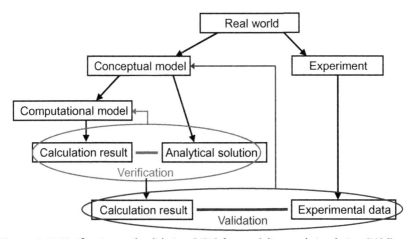

Figure 6.16 Verification and validation (V&V) for modeling and simulation (M&S).

out and we can say that the computational model has been properly made from the conceptual model.

Next the calculation result is quantitatively compared with the experimental result. In this process, various uncertainties in the conceptual model are assessed. Errors and uncertainties in the experiment are also necessary for the quantitative comparison. The laws of physics involved in the conceptual model are calculated, which means that the laws excluded in the conceptual model cannot be calculated. Thus we need to compare the calculation and the experiment to assess the validity of the conceptual model. If the validation is acceptable, the chosen laws of physics are proper. Otherwise, we need to consider improvement of the chosen laws to be more sophisticated or addition of other laws of physics.

Books of V&V for M&S have been published (Roache, 1998; Coleman and Steele, 2009; Oberkampf and Roy, 2010). The American Society of Mechanical Engineers (ASME) has published a guide and a standard for V&V for M&S (ASME, 2006, 2009). The Atomic Energy Society of Japan has published a guide (AESJ, 2016).

Another type of V&V for quality management (QM) is known as well (ISO, 2015; NAFEMS, 2008; JSCES, 2017, 2015a,b). Different definitions of V&V are given in V&V for QM. The requirements in the V&V for QM are useful to enhance the credibility in the business. Both V&Vs for M&S and QM are necessary (Shiratori et al., 2013).

After V&V for M&S, we can evaluate the errors and uncertainties in the solution of the problem in the real world. Without the errors and uncertainties the simulation results cannot be used for design and decision making. Particularly, accumulation of validation is important for the practical use. Step by step V&V will expand the application of the MPS method in industries.

6.8.2 Application to Automobile Industry

Oil circulation in a gear box was analyzed using the MPS method (Fig. 6.17). Three rotating gears were assumed. The wall boundaries were represented by polygons and the oil was analyzed by moving particles. The complex three-dimensional domain was dealt with the distance function stored in space. In Fig. 6.17 the oil particles are drawn in the left side, while realistic visualization using computer graphics is given in the right side. Oil is circulated by the rotating gear, collected in the upper

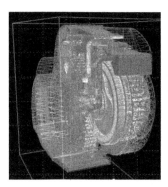

Figure 6.17 Oil circulation in an automobile gear box: particle behavior (left) and photo realistic computer graphics (right). *Source: Courtesy: Prometech Software, Inc.*

small pool and returned to the bottom pool though a pipe. The oil surface is maintained above the bottom of the lowest gear to be lifted up by the gear rotation. The MPS method is expected to analyze the oil behavior in the gear box in many situations, such as tilted, accelerated, etc.

Water behavior around a car was analyzed when it run on a flooded road (Tanaka et al., 2017). To avoid the troubles of water entry into the air intake duct and the electric devices, it is important to predict the water level around the car. Coupling between water behavior and car motion should be considered because buoyance force is effective due to the leakage tightness around the engine compartment. Thus the MPS method was coupled with a multibody dynamics code. The fluid forces acting on the rigid bodies were calculated by the MPS method and transferred to the multibody dynamics code in each time step. The car motion was calculated by the multibody dynamics code and the wall boundaries were updated in the MPS method in the next time step. The experimental and calculation results are provided in Fig. 6.18 (Tanaka et al., 2017). Water rises up in front of the car in the experiment, which is also reproduced in the calculation. It was found that there were two patterns of water behavior depending on the car speed and the flooded water depth. The calculated water level agreed with that of the experiment within 5% error.

A numerical model to analyze bubble motion in oil has been developed (Matsui et al., 2016). Bubble behavior in oil is important because the excess mixing of the air bubbles in the engine oil may cause deterioration of lubrication. The model is based on the combination of the MPS

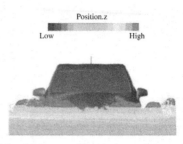

Figure 6.18 Water behavior around vehicle driving on flooded road: experiment (left) and calculation (right) (Tanaka et al., 2017). *Source: Courtesy: Toyota Motor Co.*

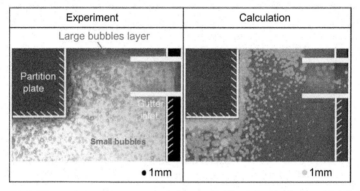

Figure 6.19 Comparison between experiment (left) and calculation (right) for air bubble behavior in the upper part of the oil tank (Matsui et al., 2016). *Source: Courtesy: Honda Motor Co., Ltd.*

method and the discrete element method (DEM). The oil behavior is analyzed by the MPS method, whereas the air bubbles are analyzed by DEM particles. The bubble transport is modeled by considering gravity, buoyancy, drag, lift, virtual mass, contact with wall, and contact among the bubbles. The contact with bubbles is evaluated by a spring and a dashpot between the DEM particles as a normal DEM modeling. Bubble disappearance occurs on the oil-free surface and is modeled by a time constant. Fig. 6.19 shows bubble behavior near the outlet (Matsui et al., 2016). The oil moves from the left to the right with many air bubbles. The outlet is located on the right side. The upper side is the free surface of the oil. Larger bubbles move upward and disappear on the free surface and small bubbles enter into the outlet. We can see good agreement between the experiment and the calculation using the bubble model developed by Matsui et al. (2016).

Yuhashi et al. (2015) carried out validation of torque of a rotating gear in a box with oil using a simple geometry. One eccentric gear was located in the center of a cylindrical box. The rotating speed was changed from 250 to 1750 rotations per minute (rpm). Fig. 6.20 shows the comparison between the experiment and the calculation at 1750 rpm. The rotation direction is anticlockwise. We can see the oil is circulated with splashing. As the rotating speed increases, more oil flows on the outer surface. The calculated torques agreed well with those of the experiment.

Grid Convergence Index (GCI) was assessed by Yuhashi et al. (2015) based on ASME V&V-20 (ASME, 2009). GCI is a procedure of verification to evaluate the error of spatial resolution. They prepared three resolutions of 2.0, 1.0, and 0.5 mm of the particle spacing using 6689, 58,858, and 477,543 particles, respectively. The order of convergence was 2.10 and the evaluated GCI was 3.04%. Difference of the torque between the converged solution and the experimental result was 1%, which was within the obtained GCI. A wide range of physical properties of oil was investigated by Yuhashi et al. (2016).

The shape of the crank-case was improved by using the Taguchi method (Fig. 6.21) (Yuhashi and Koshizuka, 2017). The Taguchi method was used with the orthogonal array L18 (Tatebayashi, 2004). Each case was simulated by the MPS method and oil flow in the target area was evaluated. The optimized shape was obtained as shown in Fig. 6.22. Compared with the initial shape, the improved one was ellipsoidal and the ribs were added on the outer surface. The shape of the products with more complicated geometry was designed by considering the present optimization result for the simplified geometry. It was reported that the oil temperature was substantially decreased in the improved design, which meant that the lubrication performance was enhanced.

Oil behavior in a commercial gear box was analyzed by the MPS method. Rotating many gears were represented by polygon walls.

Figure 6.20 Comparison of oil behavior in experiment (left) and calculation (right) in a gear box (Yuhashi et al., 2015). *Source: Courtesy: Maruyama Mfg. Co., Ltd.*

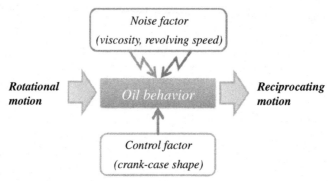

Figure 6.21 System diagram of lubrication characteristics in gear box for Taguchi method (Yuhashi and Koshizuka, 2017). *Source: Courtesy: Maruyama Mfg. Co., Ltd.*

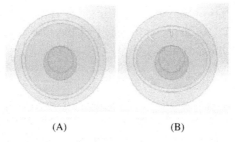

(A)   (B)

Figure 6.22 Optimized shape of gear box with simplified geometry using Taguchi method (Yuhashi and Koshizuka, 2017). (A) Initial shape; (B) optimum shape. *Source: Courtesy: Maruyama Mfg. Co., Ltd.*

Fig. 6.23 shows the oil behavior at the time of the start of the gear rotation and at 1 s after the start of the rotation. The rotation speed was 3000 rpm with the oil volume of 1950cc. The number of used particles was approximately 1.12M. The explicit algorithm using pseudo-compressibility and the potential model for surface tension were employed. We can see that oil is circulated inside the gear box. The torque of the shaft and the flowrate in the oilgutter were validated by comparing with the experiment.

Behavior of condensed water in car air conditioners is important for the performance, corrosion, and eliminating odor. For the simulation of water in the car air conditioners, Hattori et al. (2016) improved the wettability model using the potential force in the MPS method. The surface tension and wettability are modeled by an interparticle potential force

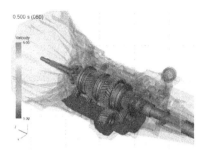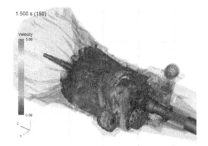

Figure 6.23 Oil behavior analysis in a commercial gear box with the rotation speed of 3000 rpm and the oil volume of 1950cc: 0 s (left) and 1 s (right) after start of rotation. *Source: Courtesy: Univance Co.*

between neighboring particles as explained in Section 5.2. When the wall boundaries are represented by the polygons in the MPS method, as explained in Section 4.2.3, we need to consider the treatment of the potential force between a fluid particle and polygons. Hattori et al. (2016) proposed a new formulation which was useful on the polygon to reproduce the contact angles more accurately.

6.8.3 Application to Chemical Engineering

Mixing and kneading in twin screw extruders have been analyzed by the MPS method (Fukuzawa et al., 2017). The resin is supplied as solid pellets from the inlet. This process is calculated by the DEM. The pellets are heated and melted by the shear stress. This process is evaluated by switching the DEM particles to the MPS particles. The comparison between the experiment and the calculation is depicted in Fig. 6.24. In the experiment, white pellets are supplied from the left and moved to the central region. The central region is filled with the pellets which are melted by shear heating. The melted pellets are observed by red color. The melted resin is moved to right. The space is fully filled by the resin in the central region, while the space is partly filled with the free surfaces in the right and left regions. The particle methods, DEM and MPS, are useful for both the fully filled and partly filled regions, as shown in Fig. 6.24. High viscous non–Newtonian fluid behavior was verified to use the MPS method for the resin in the extruders by Fukuzawa et al. (2014).

Devolatilization in mixing tanks has been modeled in the MPS method (Matsunaga et al., 2016; Hirayama et al., 2016, 2017). Each MPS

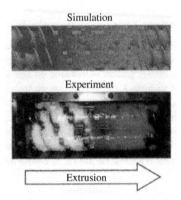

Figure 6.24 Mixing and kneading in an extruder; solid pellets are white (gray in print version) (calculated by DEM) and melted resin is red (dark gray in print version) (calculated by MPS). *Source: Courtesy: The Japan Steel Works, Ltd.*

particle representing the liquid in the mixing tanks holds new variables of number of bubbles, mole of bubbles and concentration of dissolved gas. Transport equations of these new variables are formulated and solved. Initially, gas is dissolved in a liquid in a mixing tank. Bubbles are generated and grow in the liquid and finally removed from the free surface. These processes are modeled by the increase of the number of bubbles and transfer from the dissolved gas to the mole of bubbles, respectively. It this model the size of bubbles are assumed to be smaller than the MPS particles, and volume expansion of the MPS particles is neglected. The bubbles are also assumed to move mostly together with the liquid because of high viscosity. Fig. 6.25 shows a calculation example using anchor blades with a rotation speed of 60 rpm. We can see that the bubbles are generated on the inner surface of the tank and the blade surface. Then the bubbles move to the free surface with the liquid flow induced by the blade motion. The concentration of the dissolved gas is decreased at 1800 s. The gas removal speed is roughly the same as observed in the experiment.

The fluid behavior in a mixing tank used in the cosmetic industry has been analyzed (Awasaki et al., 2014; Tanaka et al., 2014). Fig. 6.26 provides an example of the mixing tanks used for emulsification. There are three types of blades, anchor, disper, and homogenizing mixer, in the same tank. The simulation was carried out with 129,269 particles for 3.6 L fluid with a rotation speed of 90 rpm. The comparison between the simulation and the experiment is shown in Fig. 6.27. Violent motion of

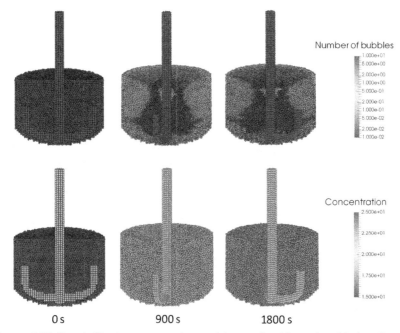

Figure 6.25 Devolatilization process in a mixing tank with anchor blades. *Source: Courtesy: Mitsubishi Chemical Co.*

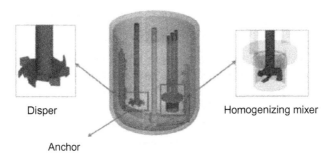

Figure 6.26 Mixing tank for emulsification. *Source: Courtesy: Shiseido Co., Ltd.*

the free surfaces is successfully reproduced in the simulation. The shear rate, which is experienced in the fluid, is the key parameter for emulsification. A model of changing the liquid viscosity due to the experienced shear rate was developed to simulate the emulsification process (Tanaka et al., 2014).

Simulation

Experiment

Figure 6.27 Comparison of experiment and calculation for emulsification process in the mixing tank. *Source: Courtesy: Shiseido Co., Ltd.*

The press process of carbon fiber reinforced thermoplastics (CFRTP) is analyzed by the MPS method (Shino et al., 2016). The matrix thermoplastics is melted by increasing the temperature during the press molding. In this study the melted CFRTP is modeled as a nonisotropic fluid represented by a matrix of viscosity in the Navier—Stokes equations

$$\frac{D\boldsymbol{u}}{Dt} = -\frac{1}{\rho}\nabla P + \boldsymbol{C}\nabla^2\boldsymbol{u} + \frac{\boldsymbol{f}}{\rho},\tag{6.44}$$

where

$$\boldsymbol{C} = \boldsymbol{\Psi}^T \begin{pmatrix} \nu_{\mathrm{hor}} & 0 & 0 \\ 0 & \nu_{\mathrm{ver}} & 0 \\ 0 & 0 & \nu_{\mathrm{ver}} \end{pmatrix} \boldsymbol{\Psi}.\tag{6.45}$$

Here, ν_{hor} and ν_{ver} are the viscosities in the horizontal and vertical directions of the fibers, respectively. The fluid is much more difficult to flow in the horizontal direction, which is modeled by $\nu_{\mathrm{ver}} < < \nu_{\mathrm{hor}}$. $\boldsymbol{\Psi}$ is the matrix of the coordinate transformation from the global coordinate system to the local coordinate system where the fiber direction is set x-coordinate. Nonisotropic spreading of CFRTP is calculated by setting, for example $\nu_{\mathrm{hor}} = 1000\nu_{\mathrm{ver}}$ (Shino et al., 2016).

6.8.4 Application to Metal Engineering

In the continuous casting process, molten steel is solidified by water spray. Uniform cooling is favorable to keep the quality of steel. Fig. 6.28 illustrates the geometry of the validation experiment of spray cooling of the

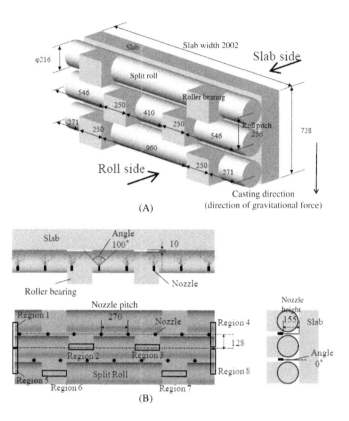

Figure 6.28 Geometry of validation calculation for spray water cooling of steel slab in continuous casting process (unit mm) (Yamasaki, 2014). (A) Perspective view of the model. (B) Three orthographic views of the model. *Source: Courtesy: Nippon Steel & Sumitomo Metal Co.*

steel slab which is supported by the rolls (Yamasaki, 2014; Yamasaki et al., 2015). The spray nozzles are located between the rolls. The rolls are separated to drain water and the flowrate of the drained water of each position is validated by comparing the calculation result and the experimental data. Good agreement is obtained (Yamasaki, 2014; Yamasaki et al., 2015). Fig. 6.29 shows an example of comparison between the MPS calculation and the experiment. We can see that water is accumulated on the lower central roll. This may cause local over-cooling of the steel slab. The mechanism of nonuniform heat transfer on the steel slab is clarified and the manufacturing process is improved.

A casting process of aluminum was analyzed using the MPS method by Regmi et al. (2015), where semisolid physical properties were

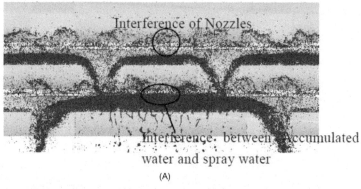

Interference of Nozzles

Interference. between accumulated water and spray water

(A)

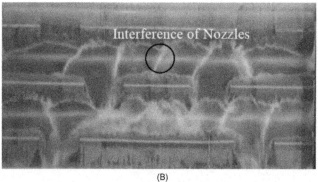

Interference of Nozzles

(B)

Figure 6.29 Calculation and experimental results of spray water cooling, view from the slab side (Yamasaki, 2014). (A) Calculated result (20 L/min/nozzle). (B) Experimental result (20 L/min/nozzle). *Source: Courtesy: Nippon Steel & Sumitomo Metal Co.*

considered. Analysis of a welding process was carried out by Saso et al. (2016), where phase change between liquid and solid, surface tension, and Marangoni force were considered.

6.8.5 Application to Biomechanics

Failure of swallowing may cause the entry of food bolus into the lungs and aspiration pneumonia. This is dangerous for elderly. The swallowing has been simulated as the coupled dynamics of the human body and the food bolus. The human body was first modeled by the combination of rigid bodies which were forced to move, and the food bolus, which is water, was modeled by fluid particles of the MPS method (Kamiya et al.,

2013a,b). Water motion was validated by the comparison with video-fluorography images of a patient in the case of successful swallowing (Kamiya et al., 2013b). Good agreement was obtained. Fig. 6.30 shows the simulation result of the successful swallowing of water.

A more advanced model is developed for the swallowing simulation. The human body was represented by a nonlinear elastic solid obeying a Moony-Rivlin constitutive law (Kikuchi et al., 2014, 2015). The elastic solid was analyzed by the Hamiltonian MPS method. The control points in the solid were forced to move to mimic the swallowing. The modeled organs related to the swallowing is shown in Fig. 6.31 and the calculation result are shown in Fig. 6.32 (Kikuchi et al., 2017). A half of the geometry is removed to see the inside. Water is initially contained above the tongue and carried into the throat. At this timing the epiglottis falls down and closes the entry to the trachea. Then, water flows into the esophagus. The successful swallowing process is validated. Unsuccessful swallowing, aspiration, was simulated by inadequate motion of the human organs. In the simulation, we can test various conditions of unusual swallowing. The simulation is expected to use for the guidance of the treatment as well as the study of the aspiration mechanics (Kikuchi, 2017).

Other biomechanics problems have been analyzed using the MPS method since it is expected to be useful for the soft materials which suffer large deformation. Lung deformation due to breathing was analyzed (Ito et al., 2010a,b, 2011). The lung shape was made from the computer

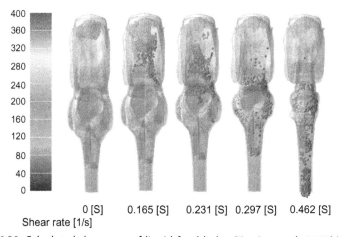

Figure 6.30 Calculated share rate of liquid food bolus (Kamiya et al., 2013b). *Source: Courtesy: Meiji Co., Ltd.*

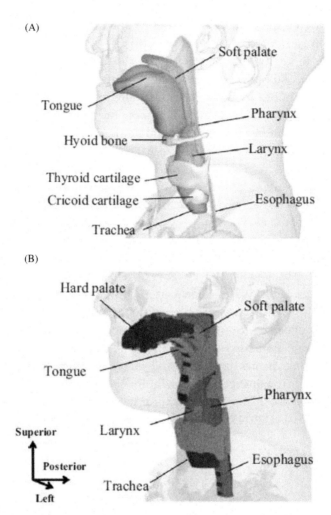

Figure 6.31 Model of organs related to swallowing (Kikuchi et al., 2017). (A) Anatomical chart; (B) Particle model. *Source: Courtesy of Japanese Red Cross Musashino Hospital.*

tomography (CT) images of a patient. The lung boundary was moved as the CT images and the local motion inside the lungs was validated. The lungs were passively moved by the active motion of the diaphragm and the ribs. The motion models of the diaphragm and ribs were developed (Ookura et al., 2013). These studies are expected to apply to the radiation therapy of lung cancers which move by breathing during the irradiation.

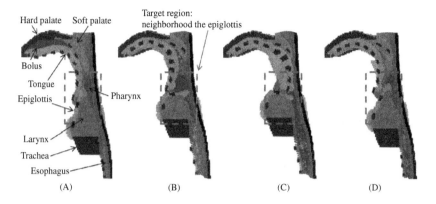

Figure 6.32 Lateral view of successful swallowing (Kikuchi et al., 2017). (A) 1.000 s; (B) 1.583 s; (C) 1.750 s; (D) 2.000 s. *Source: Courtesy: Japanese Red Cross Musashino Hospital.*

Large deformation of the breast was analyzed by Shino et al. (2013). Human posture is different for ultrasound (face-up) and magnetic resonance imaging (face-down) for detecting the breast cancers. The simulation is expected to identify the same position in the breast of different postures.

REFERENCES

Arai, J., Koshizuka, S., 2009. Numerical analysis of droplet impingement on pipe inner surface using a particle method. J. Power Energy Syst. 3, 228–236.

Arai, J., Koshizuka, S., Murozono, K., 2013. Large eddy simulation and a simple wall model for turbulent flow calculation by a particle method. Int. J. Numer. Methods Fluids 71, 772–787.

Awasaki, T., Yamada, T., Tanaka, A. and Yokokawa, Y., 2014. Effect of particle radius in unsteady mixing analysis of newtonian fluid using MPS method. In: 46th Automn Meeting on the Society of Chemical Engineering, Japan, Fukuoka, September 17–19, 2014, B215 (in Japanese).

Chai, P., Kondo, M., Erkan, N., Okamoto, K., 2016. Numerical simulation of 2D ablation profile in CCI-2 experiment by moving particle semi-implicit method. Nucl. Eng. Des. 301, 15–23.

Chen, R., Oka, Y., 2014. Numerical analysis of freezing controlled penetration behavior of the Molten Core debris in an instrument tube with MPS. Ann. Nucl. Energy 71, 322–332.

Chen, R., Tian, W., Su, G.H., Qiu, S., Ishiwatari, Y., Oka, Y., 2010. Numerical investigation on bubble dynamics during flow boiling using moving particle semi-implicit method. Nucl. Eng. Des. 240, 3830–3840.

Chen, R., Tian, W.X., Su, G.H., Qiu, S.Z., Ishiwatari, Y., Oka, Y., 2011. Numerical investigation on coalescence of bubble pairs rising in a stagnant liquid. Chem. Eng. Sci. 66, 5055–5063.

Chen, R., Oka, Y., Li, G., Matsuura, T., 2014. Numerical investigation on melt freezing behavior in a tube by MPS method. Nucl. Eng. Des. 273, 440–448.

Chen, X., Sun, Z.-G., Liu, L., Xi, G., 2016. Improved MPS method with variable-size particles. Int. J. Numer. Methods Fluids 80, 358−374.

Coleman, H.W., Steele, W.G., 2009. Experimentation, validation, and uncertainty analysis for engineering. John Wiley & Sons, Hoboken, NJ.

Duan, G., Chen, B., 2015. Large eddy simulation by particle method coupled with sub-particle-scale model and application to mixing layer flow. Appl. Math. Modell. 39, 3135−3149.

Duan, G., Chen, B., Koshizuka, S., Xiang, H., 2017. Stable multiphase moving particle semi-implicit method for incompressible interfacial flow. Comput. Methods Appl. Mech. Eng. 318, 636−666.

Duan, R.-Q., Koshizuka, S., Oka, Y., 2003a. Numerical and theoretical investigation of effect of density ratio on the critical weber number of droplet breakup. J. Nucl. Sci. Technol. 40, 501−508.

Duan, R.-Q., Koshizuka, S., Oka, Y., 2003b. Two-dimensional simulation of drop deformation and breakup at around the critical weber number. Nucl. Eng. Des. 225, 37−48.

Duan, R.-Q., Jiang, S.-Y., Koshizuka, S., Oka, Y., Yamaguchi, A., 2006. Direct simulation of flashing liquid jets using the MPS method. Int. J. Heat Mass Transfer 49, 402−405.

Erkan, N., Kawakami, T., Madokoro, H., Chai, P., Ishiwatari, Y., Okamoto, K., 2015. Numerical simulation of droplet deposition onto a liquid film by VOF-MPS method. J. Visualiz. 18, 381−391.

Fukuzawa, Y., Tomiyama, H., Shibata, K. and Koshizuka, S., 2014. Numerical analysis of high viscous non-Newtonian fluid flow using the MPS method. In: Trans. Japan Society for Computational Engineering and Science, Paper No. 20140007 (in Japanese).

Fukuzawa, Y., Shigeichi, T., Munemasa, K., Tomiyama, H., Yamanoi, M. and Koshizuka, S., 2017. Analysis of polymer plasticization by DEM-MPS coupling simulation. In: 5th Int. Conf. on Particle-based Methods. Fundamentals and Applications (Particles 2017), Hannover, September 26−28, 2017.

Ghia, U., Ghia, K.N., Shin, C.T., 1982. High-re solutions for incompressible flow using the Navier−Stokes equations and a multigrid method. J. Comput. Phys. 48, 387−411.

Gotoh, H., Khayyar, A., 2016. Current achievements and future persprctives for projection-based particle methods with applications in ocean engineering. J. Ocean Eng. Mar. Energy 2, 251−278.

Gotoh, H., Sakai, T., Shibahara, T., 2000. Lagrangian flow simulation with sub-particle-scale turbulence model. Ann. J. Hydraulic Eng. JSCE 44, 575−580 (in Japanese).

Gotoh, H., Shibahara, T., Sakai, T., 2001. Sub-particle-scale turbulence model for the MPS method. Comput. Fluid Dyn. J. 9, 339−347.

Gotoh, H., Hayashi, M., Memita, T., Sakai, T., 2003. Numerical simulation of wave overtopping on a vertical seawall by using MPS method. J. Japan Soc. Civ. Eng. 726/II-62, 87−98 (in Japanese).

Gotoh, H., Khayyar, A., Hori, C., 2009. New assessment criterion of free surface for stabilizing pressure field in particle method. J. Jpn. Soc. Civ. Eng. Ser. B2-Coast. Eng. 65, 21−25 (in Japanese).

Gotoh, H., Hori, C., Ikari, H., Khayyar, A., 2010. Semi-implicit algorithm of particle method accelerated by GPU. J. Jpn. Soc. Civ. Eng. Ser B2 (66), 217−222 (in Japanese).

Harada, T., Suzuki, Y., Koshizuka, S., Arakawa, T., Shoji, S., 2005. Simulation of droplet generation in micro flow using MPS method. Trans. Jpn. Soc. Mech. Eng. B 71, 2637−2641 (in Japanese).

Harada, T., Koshizuka, S. and Kawaguchi, Y., 2007a. A slice data structure for particle-based simulations on GPUs. In: Trans. Japan Society for Computational Engineering and Science, Paper No. 20070028 (in Japanese).

Harada, T., Tanaka, M., Koshizuka, S., Kawaguchi, Y., 2007b. Parallelization of particle-based simulations. IPSJ J. 48, 3557−3567 (in Japanese).

Harada, T., Tanaka, M., Koshizuka, S. and Kawaguchi, Y., 2007c. Real-time particle-based simulation on GPUs. In: DVD Publication at ACM SIGGRAPH Posters, San Diego, August 5−9, 2007, sap-151.

Harada, T., Masaie, I., Koshizuka, S. and Kawaguchi, Y., 2008a. Accelerating particle-based simulation utilizing spatial locality on the GPU. In: Trans. Japan Society for Computational Engineering and Science, Paper No. 20080016 (in Japanese).

Harada, T., Masaie, I., Koshizuka, S., Kawaguchi, Y., 2008b. Parallelizing particle-based simulation on multiple GPUs. IPSJ J. 49, 4067−4079 (in Japanese).

Harada, T., Masaie, I., Koshizuka, S. and Kawaguchi, Y., 2008c. Massive particles: particle-based simulations on multiple GPUs. In: ACM SIGGRAPH Talks, Los Angeles, August 11−15, 2008.

Hattori, T., Hiai, D., Akaike, S., Koshizuka, S., 2016. Improvement of wetting calculation model on polygon wall in the MPS method. Trans. Jpn. Soc. Mech. Eng. 82 (835), (in Japanese).

Heo, S., Koshizuka, S., Oka, Y., 2002. Numerical analysis of boiling on high heat-flux and high subcooling condition using MPS-MAFL. Int. J. Heat Mass Transfer 45, 2633−2642.

Hibi, S., Yabushita, K., 2004. A study on reduction of unusual pressure fluctuation of MPS method. J. Kansai Soc. Naval Arch. Jpn. 241, 125−131 (in Japanese).

Hirayama, K., Matsunaga, T., Koshizuka, S., Maki, A., Ishiba, Y., Horiguchi, A., Kikuchi, Y. and Kujime, M., 2016. Simulation of the devolatilization process using MPS method: analysis in stirred vessels. In: Proc. JSME 29th Computational Mechanics Division Conference, Nagoya, September 22−24, 2016, 069 (in Japanese).

Hirayama, K., Matsunaga, T., Koshizuka, S., Maki, A., Ishiba, Y., Horiguchi, A., Kikuchi Y. and Kujime, M., 2017. The improvement of a simulation method for the devolatilization process in stirred vessels using the MPS method. In: Proc. Conf. Computational Engineering and Science, Saitama, May 31−June 2, 2017, A-10-4 (in Japanese).

Horie, H., Shirakawa, N., Tobita, Y., Morita, K., Kondo, S., 2001. The effect of bubble size on the radial distribution of void fraction in two-phase flow in a circular tube. J. Nucl. Sci. Technol. 38, 711−720.

Hu, F., Matsunaga, T., Tamai, T., Koshizuka, S., 2017. An ALE particle method using upwind interpolation. Comput. Fluids 145, 21−36.

Ikeda, H., 1999. Numerical analysis of fragmentation processes in vapor explosions using particle method. Ph.D. Thesis, The University of Tokyo (in Japanese).

Ikeda, H., Matsuura, F., Koshizuka, S., Oka, Y., 1998. Numerical analysis of fragmentation processes of liquid metal in vapor explosions using moving particle semi-implicit method". Trans. Jpn. Soc. Mech. Eng. B 64, 2431−2437 (in Japanese).

Ikeda, H., Koshizuka, S., Oka, Y., Park, H.S., Sugimoto, J., 2001. Numerical analysis of jet injection behavior for fuel-coolant interaction using particle method. J. Nucl. Sci. Technol. 38, 174−182.

Ikejiri, S., Liu, J., Oka, Y., 2007. Simulation of a single bubble rising with hybrid particle-mesh method. J. Nucl. Sci. Technol. 44, 886−893.

Iribe, T., Nakaza, E., 2010. A precise calculation method of the gradient operator in numerical computation with the MPS. J. Jpn. Soc. Civ. Eng. Ser.B2-Coast. Eng. 66, 46−50 (in Japanese).

Iribe, T., Fujisawa, T., Shibata, K. and Koshizuka, S., 2006. Study on parallel computation for fluid simulation using MPS method. In: Trans. Japan Society for Computational Engineering and Science, Paper No. 20060015 (in Japanese).

Iribe, T., Fujisawa, T. and Koshizuka, S., 2008. Reduction of communication between nodes on large-scale simulation of the particle method. In: Trans. Japan Society for Computational Engineering and Science, Paper No. 20080020 (in Japanese).

Iribe, T., Fujisawa, T., Koshizuka, S., 2010. Reduction of communication in parallel computing of particle method for flow simulation of seaside areas. Coastal Eng. J. 52, 287–304.

Ishii, E., Ishikawa, T., Tanabe, T., 2006. Hybrid particle/grid method for predicting motion of micro- and macrofree surfaces. J. Fluids Eng. 128, 921–930.

Ishii, E., Ishikawa, M., Sukegara, Y., Yamada, H., 2011. Secondary-drop-breakup simulation integrated with fuel-breakup simulation near injector outlet. J. Fluids Eng. 133, 081302.

Ishii, E., Ehara, H., Abe, M., Ishikawa, T., 2014. Short spray penetration for direct injection gasoline engines with secondary-drop-breakup simulation integrated with fuel-breakup simulation. J. Eng. Gas Turbines Power 136, 09506.

Ito, H., Koshizuka, S., Nakagawa, K., Haga, A., 2010a. Particle simulation of lung deformation in axial plane by chest respiration and heatbeat. Trans. Jpn. Soc. Simul. Technol. 2, 93–100 (in Japanese).

Ito, H., Koshizuka, S., Shino, R., Haga, A., Yamashita, H., Onoe, T., Nakagawa, K., 2010b. and Nakagawa K., Generation method for simulation-based chest 4DCT based on particle simulation for radiotherapy. Med. Imaging Technol. 28, 229–236 (in Japanese).

Ito, H., Koshizuka, S., Haga, A., Nakagawa, K., 2011. Rib cage motion model construction based on patient-specific CT images between inhalation and exhalation. Med. Imaging Technol. 29, 208–214 (in Japanese).

Jeong, S.-M., Sato, T., Chen, B., Tabeta, S., 2011. Development of a multi-scale ocean model by using particle Laplacian method for anisotropic mass transfer. Int. J. Num. Methods Fluids 66, 49–63.

Jeong, S.-M., Nam, J.-W., Hwang, S.-C., Park, J.-C., Kim, M.-H., 2013. Numerical prediction of oil amount leaked from a damaged tank using two-dimensional moving particle simulation method. Ocean Eng. 69, 70–78.

Kamada, H., Tsubota, K., Nakamura, M., Wada, S., Ishikawa, T., Yamaguchi, T., 2010. A three-dimensional particle simulation of the formation and collapse of a primary thrombus. Int. J. Num. Methods Biomed. Eng. 26, 488–500.

Kamiya, T., Toyama, Y., Michiwaki, Y. and Kikuchi, T., 2013a. Development of a numerical simulator of human swallowing using a particle method. Part 1. Preliminary evaluation of the possibility of numerical simulation using the MPS method). In: Proc. 35th Annual Int. Conf. IEEE EMBS, Osaka, July 3–7, 2013, pp. 4454–3357.

Kamiya, T., Toyama, Y., Michiwaki, Y. and Kikuchi, T., 2013b. Development of a numerical simulator of human swallowing using a particle method. Part 2. Evaluation of the accuracy of a swallowing simulation using the 3D MPS method. In: Proc. 35th Annual Int. Conf. IEEE EMBS, Osaka, July 3–7, 2013, pp. 2992–2995.

Kawahara, T., Oka, Y., 2012. Ex-vessel Molten Core solidification behavior by moving particle semi-implicit method. J. Nucl. Sci. Technol. 49, 1156–1164.

Khayyer, A., Gotoh, H., 2011. Enhancement of stability and accuracy of the moving particle semi-implicit method. J. Comput. Phys. 230, 3093–3118.

Kikuchi, T., 2017. Study on swallowing simulation considering interaction between human body and food borus. Ph.D. Thesis, The University of Tokyo (in Japanese).

Kikuchi, T., Michiwaki, Y., Koshizuka, S., Kamiya, T., Osada, T., Jinno, N. and Toyama, Y., 2014. Simulation of uniaxial compression based on Hamiltonian MPS method with wal boundary condition using penalty method. In: Trans. Japan Society for Computational Engineering and Science, Paper No. 20140010 (in Japanese).

Kikuchi, T., Michiwaki, Y., Kamiya, T., Toyama, Y., Tamai, T., Koshizuka, S., 2015. Human swallowing simulation based on videofluorography images using hamiltonian MPS method. Comput. Part. Mech. 2, 247−260.

Kikuchi, T., Michiwaki, Y., Koshizuka, S., Kamiya, T., Toyama, Y., 2017. Numerical simulation of interaction between organs and food bolus during swallowing and aspiration. Comput. Biol. Med. 80, 114−123.

Kondo, M. and Koshizuka, S., 2008. Suppressing the numerical oscillation in moving particle semi-implicit method. In: Trans. Japan Society for Computational Engineering and Science, Paper No. 20080015 (in Japanese).

Kondo, M., Koshizuka, S., 2011. Improvement of stability in moving particle semi-implicit method. Int. J. Num. Methods Fluids 65, 638−654.

Koshizuka, S., Ikeda, H., Oka, Y., 1999. Numerical analysis of fragmentation mechanisms in vapor explosions. Nucl. Eng. Des. 189, 423−433.

Lee, B.-H., Park, J.-C., Kim, M.-H., Hwang, S.-C., 2011. Step-by-step improvement of MPS method in simulating violent free-surface motions and impact-loads. Comput. Methods Appl. Mech. Eng. 200, 1113−1125.

Lele, S.K., 1992. Compact finite difference schemes with spectal-like resolution. J. Comput. Phys. 103, 16−42.

Li, G., Oka, Y., Furuya, M., 2014. Experimental and numerical study of stratification and solidification/melting behavior. Nucl. Eng. Des. 272, 109−117.

Li, G., Liu, N., Duan, G., Chong, D., Yan, J., 2016. Numerical investigation of erosion and heat transfer characteristics of molten jet impingement onto solid plate with MPS-LES method. Int. J. Heat Mass Transfer 99, 44−52.

Li, X., Oka, Y., 2014. Numerical simulation of the SURC-2 and SURC-4 MCCI experiments by MPS method. Ann. Nucl. Energy 73, 46−52.

Li, X., Yamaji, A., 2017. Three-dimensional numerical study on the mechanism of anisotropic MCCI by improved MPS method. Nucl. Eng. Des. 314, 207−216.

Liu, J., Koshizuka, S., Oka, Y., 2005. A hybrid particle-mesh method for viscous, incompressible, multiphase flows. J. Comput. Phys. 202, 65−93.

Matsui, K., Matsushita, K., Ri, H., Masaie, I., Yamanoi, M., 2016. Studies on particle method simulation of bubble behavior in engine lubricating oil (furst report). In: 2016 JSAE Annual Congress (Spring), Yokohama, May 25−27, 2016, 20165397, pp. 2119−2124 (in Japanese).

Matsunaga, T., Koshizuka, S., Maki, A., Ishiba, Y., Horiguchi, A., Kikuchi, Y., and Kujime, M., 2016. Simulation of the devolatilization process using MPS method: development of bubble model. In: Proc. JSME 29th Computational Mechanics Division Conference, Nagoya, September 22−24, 2016, 068 (in Japanese).

Morita, K., Zhang, S., Koshizuka, S., Tobita, Y., Yamano, H., Shirakawa, N., Inoue, F., Yugo, H., Naito, M., Okada, H., Yamamoto, Y., Himi, M., Hirano, E., Shimizu, S., Oue, M., 2011. Detailed analysis of key phenomena in core disruptive accidents of sodium-cooled fast reactors by the COMPASS code. Nucl. Eng. Des. 241, 4672−4681.

Murotani, K., Oochi, M., Fujisawa, T., Koshizuka, S. and Yoshimura, S., 2012. Distributed memory parallel algorithm for explicit MPS using ParMETIS. In: Trans. Japan Society for Computational Engineering and Science, Paper No. 20120012 (in Japanese).

Murotani, K., Koshizuka, S., Tamai, T., Shibata, K., Mitsume, N., Yoshimura, S., Tanaka, S., Hasegawa, K., Nagai, E., Fujisawa, T., 2014. Development of hierarchical domain decomposition explicit MPS method and application to large-scale tsunami analysis with floating objects. J. Adv. Simul. Sci. Eng. 1, 16−35.

NAFEMS. 2008. Engineering simulation — quality management systems — requirements. NAFEMS QSS:2008.

Nomura, K., Koshizuka, S., Oka, Y., Obata, H., 2001. Numerical analysis of droplet breakup behavior using particle method. J. Nucl. Sci. Technol. 38, 1057–1064.

Oberkampf, W.L., Roy, C.J., 2010. Verification and Validation in Scientific Computing. Cambridge University Press, Cambridge, UK.

Ookura, T., Ito, H., Koshizuka, S., Nomoto, A., Haga, A., Nakagawa, K., 2013. Diaphragm respiratory motion model with ribcage movement. Med. Imaging Technol. 31, 189–197 (in Japanese).

Regmi, A., Shintaku, H., Sasaki, T., Koshizuka, S., 2015. Flow simulation and solidification phenomena of AC4CH aluminum alloy in semi-solid forging process by explicit MPS method. Comput. Part. Mech. 2, 223–232.

Roache, P.J., 1998. Verification and validation in computational science and engineering. Hermosa.

Saso, S., Mouri, M., Tanaka, M., Koshizuka, S., 2016. Numerical analysis of two-dimensional welding process using particle method. Weld. World 60, 127–136.

Shibata, K., Koshizuka, S., Oka, Y., 2004. Numerical analysis of jet breakup behavior using particle method. J. Nucl. Sci. Technol. 41, 715–722.

Shibata, K., Masaie, I., Kondo, M., Murotani, K., Koshizuka, S., 2015. Improved pressure calculation for the moving particle semi-implicit method. Comput. Part. Mech. 2, 91–108.

Shibata, K., Koshizuka, S., Masaie, I., 2016. Cost reduction of particle simulations by an ellipsoidal particle model. Comput. Methods Appl. Mech. Eng. 307, 411–450.

Shibata, K., Koshizuka, S., Matsunaga, T., Masaie, I., 2017. The overlapping particle technique for multi-resolution simulation of particle methods. Comput. Methods Appl. Mech. Eng. 325, 434–462.

Shino, R., Koshizuka, S., Ito, H., Jibiki, T., Liu, L., 2013. Generation of a supine image of the breast using numerical simulation of the particle method from a prone image. Med. Imaging Technol. 31, 240–247 (in Japanese).

Shino, R., Tamai, T., Koshizuka, S., Maki, A. and Ishikawa, K., 2016. Anisotropic high viscosity fluid analysis using a particle method for evaluating CFRTP press molding process. In: Trans. Japan Society for Computational Engineering and Science, Paper No. 20160015 (in Japanese).

Shirakawa, N., Horie, H., Yamamoto, Y., Tsunoyama, S., 2001a. Analysis of the void distribution in a circular tube with the two-fluid particle interaction method. J. Nucl. Sci. Technol. 38, 392–402.

Shirakawa, N., Horie, H., Yamamoto, Y., Okano, Y., Yamaguchi, A., 2001b. Analysis of jet flows with the two-fluid particle interaction method. J. Nucl. Sci. Technol. 38, 729–738.

Shirakawa, N., Yamamoto, Y., Horie, H., Tsunoyama, S., 2003. Analysis of subcooled boiling with the two-fluid particle interaction method. J. Nucl. Sci. Technol. 40, 125–135.

Shiratori, M., Koshizuka, S., Yoshida, Y., Nakamura, H., Hotta, A. and Takano, N., 2013. Quality assurance and V&V for engineering simulation. Maruzen Shuppan (in Japanese).

Sueyoshi, M., 2005. Validation of a numerical prediction method of impulsive pressure by particle method. In: Proc. 18th Ocean Engineering Symposium, June 27–28, 2005 (in Japanese).

Sugimoto, K., Okauchi, K., Zannino, D., Brizard, C.P., Liang, F., Sugawara, M., Liu, H., Tsubota, K., 2015. Total cavopulmonary connection is superior to atriopulmonary connection fontan in preventing thrombus formation: computer simulation of flow-related blood coagulation. Pediatr. Cardiol. 36, 1436–1441.

Sun, Z., Xi, G., Chen, X., 2009. Mechanism study of deformation and mass transfer for binary droplet collisions with particle method. Phys. Fluid 21, 032106.

Suzuki, Y., 2007. Higher-order particle method and multi-physics simulator. Ph.D. Thesis, The University of Tokyo (in Japanese).

Suzuki, Y., Koishikawa, M., Koshizuka, S., Okamoto, T., Kaneko, N., Takamatsu, A., Fujii, T., 2006. Numerical simulation of adhesion of cells in micro channels using the MPS method. Trans. Jpn. Soc. Mech. Eng. B 72, 2109−2116 (in Japanese).

Suzuki, Y., Koshizuka, S., Oka, Y., 2007. Hamiltonian moving-particle semi-implicit (HMPS) method for incompressible fluid flows. Comput. Methods Appl. Mech. Eng. 196, 2876−2894.

The American Society of Mechanical Engineers (ASME), 2006. Guide for verification and validation in computational solid mechanics. ASME V&V 10-2006.

The American Society of Mechanical Engineers (ASME), 2009. Standard for verification and validation in computational fluid dynamics and heat transfer. ASME V&V 20-2009.

The Japan Society for Computational Engineering and Science (JSCES), 2017. Quality management of engineering simulation. JSCES S-HQC001:2017 (in Japanese).

The Japan Society for Computational Engineering and Science (JSCES), 2015a. A model procedure for engineering simulation. JSCES S-HQC002:2015 (in Japanese).

The Japan Society for Computational Engineering and Science (JSCES), 2015b. The application examples for quality management of engineering simulation (HQC001) & a model for engineering simulation (HQC002). JSCES S-HQC003:2015 (in Japanese).

Tamai, T., Koshizuka, S., 2014. Least squares moving particle semi-implicit method. Comput. Part. Mech. 1, 277−305.

Tamai, T., Shibata, K. and Koshizuka, S., 2013. Development of the higher-order MPS method using the taylor expansion. In: Trans. Japan Society for Computational Engineering and Science, Paper No. 20130003 (in Japanese).

Tanaka, A., Yokokawa, Y., Awasaki, T. and Yamada, T., 2014. Numerical analysis of emulsion using a key of shear rate with viscosity change by mixing. In: 46th Automn Meeting on the Society of Chemical Engineering, Japan, Fukuoka, September 17−19, 2014, B216 (in Japanese).

Tanaka, M., Masunaga, T., 2010. Stabilization and smoothing of pressure in MPS method by quasi-compressibility. J. Comput. Phys. 229, 4279−4290.

Tanaka, M., Masunaga, T., Nakagawa, Y., 2009. Multi-resolution MPS method. In: Trans. Japan Society for Computational Engineering and Science, Paper No. 20090001 (in Japanese).

Tanaka, Y., Yamamura, J., Murakawa, A., Tanaka, H., Yasuki, T., 2017. Development of prediction method for engine compartment water level by using coupled mutibody and fluid dynamics. SAE Int. J. Passeng. Cars-Mech. Syst. 10, 2017-01-1328.

Tang, Z., Wan, D., Chen, G., Xiao, Q., 2016. Numerical simulation of 3D violent free-surface flows by multi-resolution MPS method. J. Ocean Eng. Mar. Energy 2, 355−364.

Tatebayashi, K., 2004. Introduction to Taguchi Methods. JUSE Press, Shibuya-Ku, Japan (in Japanese).

The Atomic Enegry Society of Japan (AESJ), 2016. Guideline for credibility assessment of nuclear simulations: 2015. AESJ-SC-A008:2015 (in Japanese).

The International Organization for Standardization (ISO), 2015. Quality management systems − requirements. ISO 9001:2015.

Tian, W., Ishiwatarim, Y., Ikejiri, S., Yamakawa, M., Oka, Y., 2009. Numerical simulation on void bubble dynamics using moving particle semi-implicit method. Nucl. Eng. Des. 239, 2382−2390.

Tian, W., Qui, S.-Z., Su, S.-H., Ishiwatari, Y., Oka, Y., 2010a. Numerical solution on spherical vacuum bubble collapse using MPS method. J. Eng. Gas Turbines Power 132, 102920.

Tian, W., Ishiwatari, Y., Ikejiri, S., Yamakawa, M., Oka, Y., 2010b. Numerical computation of thermally controlled steam bubble condensation using moving particle semi-implicit (MPS) method. Ann. Nucl. Energy 37, 5−15.

Tsukamoto, M.M., Cheng, K.-Y., Motezuki, F.K., 2016. Fluid interface detection technique based on neighborhood particles centroid deviation (NPCD) for particle method. Int. J. Num. Methods Fluids 82, 148−168.

Xie, H., Koshizuka, S., Oka, Y., 2004a. Numerical simulation of liquid drop deposition in annular-mist flow regime of boiling water reactor. J. Nucl. Sci. Technol. 41, 569−578.

Xie, H., Koshizuka, S., Oka, Y., 2004b. Modelling of a single drop impact onto liquid film using particle method. Int. J. Num. Methods Fluids 45, 1009−1023.

Xie, H., Koshizuka, S., Oka, Y., 2005. Simulation of drop deposition process in annular mist flow using three-dimensional particle method. Nucl. Eng. Des. 235, 1687−1697.

Xiong, J., Koshizuka, S., Sakai, M., 2010. Numerical analysis of droplet impingement using the moving particle semi-implicit method. J. Nucl. Sci. Technol. 47, 314−321.

Xiong, J., Koshizuka, S., Sakai, M., 2011. Investigation of droplet impingement onto wet walls based on simulation using particle method. J. Nucl. Sci. Technol. 48, 145−153.

Xiong, J., Koshizuka, S., Sakai, M., Ohshima, H., 2012. Investigation on droplet impingement erosion during steam generator tube failure accident. Nucl. Eng. Des. 249, 132−139.

Yamasaki, N., 2014. Numerical analysis of spray water flow in secondary cooling of continuous casting process of steel using MPS method. In: Trans. Japan Society for Computational Engineering and Science, Paper No. 20140016 (in Japanese).

Yamasaki, N., Shima, S., Tsunenari, K., Hayashi, S., Doki, M., 2015. Particle-based numerical analysis of spray water flow in secondary cooling of continuous casting machines. ISIJ Int. 55, 976−983.

Yoon, H.Y., Koshizuka, S., Oka, Y., 1999a. A particle-gridless hybrid method for incompressible flows. Int. J. Numer. Methods Fluids 30, 407−424.

Yoon, H.Y., Koshizuka, S., Oka, Y., 1999b. Mesh-free numerical method for direct simulation of gas−liquid phase interface. Nucl. Sci. Eng. 133, 192−200.

Yoon, H.Y., Koshizuka, S., Oka, Y., 2001. Direct calculation of bubble growth, departure and rise in nucleate boiling. Int. J. Multiphase Flow 27, 277−298.

Yuhashi, N. and Koshizuka, S., 2017. Optimization of cranck-case shape by using quality engineering and moving particle semi-implicit method. In: Trans. Japan Society for Computational Engineering and Science, Paper No. 20170006 (in Japanese).

Yuhashi, N., Natsuda, I. and Koshizuka, S., 2015. Calculation and evaluation of torque generated by the rotation flow using moving particle semi-implicit method. In: Trans. Japan Society for Computational Engineering and Science, Paper No. 20150007 (in Japanese).

Yuhashi, N., Matsuda, I., Koshizuka, S., 2016. Calculation and validation of stirring resistance in cam-shaft rotation using the moving particle semi-implicit method. J. Fluid Sci. Technol. 11, JFST0018.

Yun, G., Ishiwatari, Y., Ikejiri, S., Oka, Y., 2010. Numerical analysis of the onset of droplet entrainment in annular two-phase flow by hybrid method. Ann. Nucl. Energy 37, 230−240.

INDEX

Note: Page numbers followed by "*f*" and "*t*" refer to figures and tables, respectively.

A

Acceleration, 35
  of fluid particle, 66—67
  of gravity, 31—32
  of particles, 31—33
  term, 59—61, 166, 171
Accuracy, 122—123
Adjacency, 3—5, 148
Adjacent particles, 4
Advection equation, 158—159
ALE. *See* Arbitrary Lagrangian-Eulerian
  (ALE)
American Society of Mechanical Engineers
  (ASME), 260
Anisotropic weighting function for
  Laplacian model, 257—258
Arbitrary Lagrangian-Eulerian (ALE),
  134—140, 139*f*, 238
  formulation, 156
  movement of particle in, 134*f*
Arithmetic operations, 32—33, 62
Artificial particles for continuum
  mechanics, 6—7
ASME. *See* American Society of
  Mechanical Engineers (ASME)
Aspect ratio, 3, 255—257
Assumption of vacuum, 158, 192
Atmospheric pressure, 73, 192—193, 200
A-type free-surface particles, 206—207,
  206*f*
Automobile industry application,
  260—265, 263*f*, 264*f*
  oil behavior analysis in commercial gear
    box, 265*f*
  optimized shape of gear box, 264*f*

B

Background grid, 180
Bead dynamics, 6—7
Biomechanics application, 270—273

Boiling, 229
  effect, 238
  model for MPS method, 236—237
Boundary condition(s), 73—74, 155—156,
  192—193. *See also* Dirichlet boundary
  condition
  effective domain, 158*f*
  free surface, 189—207
  inlet and outlet boundary modeling,
    207—212
  of pressure, 73—74
  solid wall boundaries, 159—188
  of velocity, 74
Boundary integral—based method,
  180—182, 188
Boundary integral—based polygon
  representation, 162, 180—188
  full effective domain, 182*f*
  geometric relations of truncated domain,
    183*f*
Boundary particle, 89, 161—163, 165,
  168—169
Breakup, 229
  behavior, 217, 229
  phenomena, 229—230
B-type free-surface particles, 206—207, 206*f*
Bubble, 138—139, 229—230, 241,
  265—266
  behavior, 261—262, 262*f*
  growth and departure, 138—139
    on heated wall, 241
    in nucleate boiling, 138*f*
    in hybrid particle mesh method,
      229—230
  motion in oil, 261—262
Bucket method, 108

C

C language program, 30, 83—85, 88
CAD. *See* Computer-aided design (CAD)

calculateGravity function, 93, 93*f*
calculateNumberDensity() function, 96,
  97*f*
calculateNZeroAndLambda() function, 90,
  90*f*
calculateParticleNumberDensity() function,
  95
calculatePressure() function, 95–96, 96*f*
calculatePressure_forExplicitMPS()
  function, 101–102, 102*f*
calculatePressureGradient() function,
  100–101, 101*f*
calculatePressureGradient_forExplicitMPS()
  function, 102, 103*f*
calculateViscosity function, 93–94, 94*f*
Carbon fiber reinforced thermoplastics
  (CFRTP), 268
Casting process of aluminum, 269–270
CCUP method, 114
Central limit theorem, 16
CFRTP. *See* Carbon fiber reinforced
  thermoplastics (CFRTP)
Channel flow, 244
Chemical engineering application,
  265–268
  mixing and kneading in extruder, 266*f*
Classic wall particle method, 167–168
Closest point search
  algorithm, 168
  on NURBS curve, 173–174
Coalescence, 240
Coefficient matrix, 72–73, 84, 98–99
Collision, 92
  distance, 201
  function, 92
  model, 201, 204
Color function, 156–157, 222–223,
  229–230
Compact scheme, 249–250, 250*f*
Compensation of particle deficiency,
  159–160
Compiler, 85
Compressibility, 98–99, 112
  artificial, 122
  of liquid, 58
Compressible flows, 21

Compressible-incompressible unified
  algorithm, 111–118, 242
Computation time, 4, 83, 123
Computational complexity, 173–174
Computational fluid dynamics, 3,
  207–208
Computational model, 259
Computer graphics, 142–143, 260–261
Computer tomography (CT), 271–272
Computer-aided design (CAD), 162
  software, 176
Conceptual particles, 204, 204*f*
Conditional equation, 68
Conjugate gradient method, 72–73, 83,
  95–96
Connection, 3–4
Conservation
  of mass, 34–35
  of momentum, 35
Conservative pressure gradient scheme,
  205
Consistent division of space, 5
Constant coefficient, 182
Contact angle, 227–228, 230–231
Continue sentence, 93–94
Continuity, 5
  equation, 3, 16–17
  in grid-based methods, 38–39
Continuous casting process, 268–269, 269*f*
Continuum mechanics, 5–11
  errors in simple rotation using particle
    method, 10*f*
  one-dimensional pure convection, 7*f*
    finite difference method for, 9*f*
    particle method for, 9*f*
  particle dynamics, 6*f*
Continuum surface force models (CSF
  models), 190–191, 217, 241. *See also*
  Surface tension models in particle
  methods
  surface tension calculation
    CSF-based model, 218–222
    surface tension models based on CSF
      equation, 222–223
Convective acceleration, 65
Convective term, 65–66

Conventional distance function–based
    method, 180–181
Correction
    factor, 172–173
    process, 114
Corrective matrix, 240, 247–248
Courant conditions, 122
Courant number, 122
CSF models. *See* Continuum surface force
    models (CSF models)
CT. *See* Computer tomography (CT)
Curvature, 37–38, 222
Cygwin compiler, 85

D

Dam breaking, 59–61, 83
Declarations, 86–87
#define sentences, 86–87
Delta function, 15–16, 222
DEM. *See* Discrete Element Method
    (DEM)
Density ratio, 191, 238
Density-invariant form, 202–203
Devolatilization
    in mixing tanks, 265–266
    process in mixing tank with anchor
    blades, 267f
Diffusion equation, 36–37, 54–55
Dirichlet boundary
    condition, 73, 97–99, 165–167, 171,
    238. *See also* Boundary condition(s)
    constant pressure on free-surface
    particles for, 201
    for pressure, 200
    of pressure Poisson equation, 95
Dirichlet pressure, 210–212
Discrete Element Method (DEM), 7, 150,
    261–262
Discretization, 32–33
Discretized equations, 3–4
Discretized governing equations, 6
Distance function,
    156–157, 180
Distance function–based method,
    180–181, 187
Distance function–based polygon
    representation, 162, 176–180

relation between wall contact force and
    position correction, 178f
virtual boundary particles and wall
    weight function, 177f
wall normal vector, 176f
Divergence, 69
    model, 121
    theorem, 142, 182–183, 185–186
    of velocity, 38, 69–70
Divergence-free form, 202–203
Domain decomposition technique, 83–84,
    252–253
Double type, 29
Droplet, 117
    behavior in microchannel, 242
    oscillation using potential model, 226f
Dummy particles, 164–165, 164f
Dummy wall particle, 89, 97, 106

E

Eddy viscosity, 243–244
Effective domain, 157–158, 158f,
    181–182, 182f, 209–210
Effective radius (r_e), 11–12, 20, 114, 116,
    148
    from fluid particle, 162–163
    of interaction zone, 41–44, 51–53,
    90–91
Elastic, 146
    beam, 149
    solid, 271
    solid model, 146–147
    strain energy, 148–149
Electromagnetic force, 5
Ellipsoidal particle model, 255–257, 256f
E-MPS. *See* Explicit MPS (E-MPS)
emps.c sample program, 74–83, 85
Emulsification process, 266–267, 267f,
    268f
Energy conservation, 113–114, 125, 149
Energy conservation equations, 113–114,
    125
Enforcement of physical boundary
    condition, 159–160
Enthalpy (H), 233–234
Equation of continuity, 32–33, 35,
    38–39, 57–58, 67–68

Equation of motion, 26−28, 31
Equation of state, 32−33, 40, 113, 115,
    117−119
Error accumulation, 246−247
Euler's equations, 144, 246−247
Eulerian boundary-tracking method,
    156−157
Eulerian description, 2, 8
Explicit algorithm, 126
    using least square approximation,
        147−148
    using pseudo-compressibility, 118−125
Explicit method, 58−59
Explicit MPS (E-MPS), 18−19, 74, 102,
    119−121, 124−125, 124f
Extended algorithms
    ALE, 134−140, 134f, 139f
    compressible-incompressible unified
        algorithm, 111−118
    explicit algorithm using pseudo-
        compressibility, 118−125
    rigid body model, 140−146, 141f
    structural analysis, 146−150
    symplectic scheme, 125−134

F

Fast solver, 54, 83
FC. *See* Fuel cell (FC)
Finite difference method (FDM), 56, 156,
    247−248
Finite element method (FEM), 1−2, 156
Finite volume method (FVM), 1−2, 156,
    245
First partial derivative, 47
First-order Taylor expansion, 249
Fixed ghost particle, 168
    interpolation points for, 167f
    technique, 166−167
Flags, 97
Fluid
    dynamics, 5, 189−190, 259
    flow simulation, 176−177
    fluid−fluid interface, 157−158
    method, 156−157
    particle, 32, 36, 67, 98−99, 244
    phenomena, 40−41
Fluid acceleration, 35

Fluid behavior in mixing tank, 266−267
Fluid contribution term, 161−162
Fluid density, 35−36, 38−40, 68−70
    relationship between particle number
        density and, 43−44
Fluid-rigid body interaction, 146
Fluid−structure interaction problems, 168
Fluid−thin structure interaction, 149
Food bolus, 270−271, 271f
for sentence of C language, 28−29, 88, 93,
    97
Forward Euler method, 63
Four basic operations of arithmetic,
    32−33, 46−47, 62
Fractional step method, 61, 68−69
Free compilers, 84−85
Free surface, 158, 189−207
    free-surface particle detection, 193−200
    pressure calculation, 200−207
Free-slip condition, 159−160, 163−164
Free-surface boundary conditions, 202
Free-surface flow, 156−158, 203
Free-surface particle(s), 73, 97, 99,
    192−193, 210
    detection, 193−200, 194f, 197f
Friction force, 36−37
Front-tracking method, 156
Fuel cell (FC), 230−231
Full explicit algorithm, 21
FVM. *See* Finite volume method (FVM)

G

Gas, 58
    phase, 191−192
    simulations, 40
Gas diffusion layer (GDL), 230−231
Gas-liquid phase change model, 236−242
Gas−liquid two-phase flow
    analysis, 249
    and phase change model, 236−242
Gauge pressure, 73
Gauss theorem, 142
Gaussian elimination, 84, 95−96, 99
Gaussian function, 15−16
gcc. *See* GNU Compiler Collection (gcc)
GCI. *See* Grid Convergence Index (GCI)
GDL. *See* Gas diffusion layer (GDL)

Geometric singularities, 168
Ghost particle, 89, 93, 97, 107
Ghost particle method. *See* Mirror particle
    method
Global interaction physics, 5
GNU Compiler Collection (gcc), 84–85
Governing equations, 35–40
    equation of continuity, 38–39
    Navier–Stokes equations, 35–38
    vector notation, 39–40
Gradient, 46
    of quantity, 47
    vector, 12–14, 13*f*
Gradient model, 13–14, 13*f*, 48–49,
    100–101, 121
    meaning of parts, 47–51
    of MPS method, 46–47, 48*f*, 50*f*
Gravity, 5, 59–61, 83, 95
Grid Convergence Index (GCI), 263
Grid method, 65–66

H

Hagen–Poiseuille flow, 165–166
Hamiltonian system, 129, 131
Hamilton's canonical equations, 125–126,
    128–129, 133–134
Heat energy, 237–238
Heat flux, 166
High viscous non-Newtonian fluid
    behavior, 265
Higher-order schemes, 247–250
    flow velocity distributions, 251*f*
    square cavity problem, 251*f*
Homogeneous Neumann boundary
    condition, 165–166, 171
Hooke's law, 259
Hyper-elastic, 147, 149

I

Identification number (ID), 27–28
Image method, 169–171
Immersed boundary method, 156
Implicit method, 58–59
Implicit pressure Poisson equation, 21
#include sentences, 86–87
Incompressibility, 11

Incompressible flow, 11, 58–59, 118–119
Incompressible semi-implicit algorithm, 22
Incompressible Smoothed Particle
    Hydrodynamics (ISPH), 18–19, 22, 140
Index function, 156–157
Industrial application, 4
Infinitesimal volume, 32
Initial distance between particles, 26, 28
Initial positions of particles, 28–29
Initial velocity
    of fluid, 29
    of particles, 29
initializeParticlePositionAndVelocity_for
    2dim() function, 88–89, 89*f*
Injection point, 209–210
Inlet boundary, 107, 158–159
    modeling, 207–212, 209*f*, 211*f*
Input files, 83
Interaction zone, 45, 105
Interface capturing method, 156–157
Internal particle, 42–43
Interpolation points, 165–166, 167*f*
ISPH. *See* Incompressible Smoothed
    Particle Hydrodynamics (ISPH)
*i*th particle, 27–28
iTimeStep, 91

J

Jet
    breakup behaviors, 229
    impingement, 238

K

Kernel, 20
Kinematic coefficient of viscosity, 32
Kinematic viscosity coefficient, 37–38

L

Lagrange derivative, 32
Lagrange multiplier, 129–132
Lagrangian derivative, 243
    operator, 61–62
Lagrangian description, 2–3, 9–10
    particles, 2–3
Lagrangian interface tracking, 156–157

Lagrangian meshless calculation algorithm, 157—158
Lagrangian method, 208
Laplacian calculation, 55—57
Laplacian model, 14—15, 15*f*, 53—55, 112—113, 116—117
  example of Laplacian calculation, 55—57
  meaning, 54—55
  of MPS method, 53—57, 54*f*, 255—257
  particle arrangement, 55*f*
Laplacian of pressure, 184—185
Laplacian of velocity, 184—185
Laplacian operator, 14—15, 32—33, 36—37, 53—55
  and uses, 53
Large density ratio, 189—191, 238
Large Eddy Simulation (LES), 229—230
LDI. *See* Liquid Droplet Impingement (LDI)
Leap-frog scheme. *See* Störmer-Verlet scheme
Least square(s), 249—250
  approximation, 147—148
  method, 160—161, 249—250
Least Squares MPS (LSMPS), 139—140, 249—250, 250*f*
LES. *See* Large Eddy Simulation (LES)
Libraries and declarations, 86—87
Linear equations, 72—73
Linear extrapolation, 171—173
Linear mapping, 126
Liquid, compressibility of, 58
Liquid Droplet Impingement (LDI), 117, 118*f*, 119*f*, 242
Liquid flows, 58—59
Liquid phase, 239
Liquid-solid phase change, 233—234
  model, 233—236, 234*f*
Liquid—gas interactions, 191—192
Look-up table, 176—177, 179
LSMPS. *See* Least Squares MPS (LSMPS)
Lung
  deformation, 271—272
  motion, 271—272

M

M&S. *See* Modeling and simulation (M&S)
Mach number, 118—119, 123

MAFL. *See* Meshless advection using flow-directional local-grid (MAFL)
main() function, 88
mainLoopOfSimulation() function, 75*t*, 91—92, 92*f*
Mass, 236—237
  conservation, 3
    equations, 113—114, 125
  distribution, 3
  flux, 207—208
  of gas particle, 236—237
  imbalance, 210
  of particle, 33—35
Material derivative, 32, 61—62
Material point, 32
math.h functions, 86—87
Mathematical Library, 85
MD. *See* Molecular dynamics (MD)
Mechanical energy conservation, 125, 129
Melting/solidification model, 230—231, 233—236
Mesh, 3—5
  distortion, 156
  generation, 4—5, 165
  mesh-based methods, 156
  methods, 1—3, 2*f*, 10
Meshless advection using flow-directional local-grid (MAFL), 135, 136*f*
Meshless discretization, 3—5, 4*f*
Metal engineering application, 268—270, 270*f*
Method of images, 169—171
Microscale method, 226
Microsoft Visual Studio Community or Express, 84
MinGW, gcc of, 85
Mirror image, 168—169, 172
Mirror particle, 168—169, 170*f*, 171—172, 172*f*, 175—176
Mirror particle method, 168—169, 175—176
Mirror particle representation, 168—176, 170*f*
  with respect to circular boundary, 172*f*
  unsuccessful examples of mirror particle generation, 174*f*

Mirroring
   process, 167—168
   rules, 168—169, 172—175
Mixing and kneading, 265, 266*f*
Mixing tank, 265—266, 267*f*, 268*f*
Modeling and simulation (M&S), 259, 259*f*
Molecular dynamics (MD), 7, 226
Molecular force, 223
Molten core concrete interaction,
   235—236
Moment matrix, 196—198
Momentum conservation equation, 5, 114,
   125, 236—237
Momentum conservation equations,
   113—114, 125
Monolithic approaches, 189—191
moveParticle() function, 95, 95*f*
Moving boundary problem, 156
Moving least squares approximation,
   166—167
Moving particle, 2—3
Moving particle semi-implicit method
   (MPS method), 5—6, 11—20, 22, 111,
   146—147, 158, 160, 168, 190—191,
   193, 217, 218*f*, 233—235, 238,
   241—245, 250—251, 260—261
   basic theory, 33—74
      approximation of partial differential
         operators, 45—57
      governing equations, 35—40
      mass of particle, 33—35
      particle number density and weight
         function, 40—45
      semi-implicit method, 57—74
   best effective radius of interaction zone,
      105
   compressed fluid, 107
   discretization schemes, 160—161
   drawbacks and strong points, 108
   dummy wall particles, 105—106
   elements, 26—33
      acceleration of particles, 31—33
      moving particles, 29—30
      particle arrangement and expressions,
         27*f*
      setting initial positions of particles,
         28—29

setting initial velocities of particles, 29
   water column collapse simulation, 26*f*
   exercise of simulation, 103—104
   hints of exercises, 104—105
   inlet or outlet boundary in simulation
      program, 107
   particle interaction models, 12—16
   particles penetrating wall, 106
   semi-implicit algorithm, 16—18
   simulation diverging, 107
   simulation programs, 74—102
   and SPH, 18—20
   surface tension models applications
      using, 229—231
   time increment, 106—107
   time-consuming part in MPS
      simulation, 108
   weighted difference, 11—12, 11*f*, 12*f*
Moving particle semi-implicit-meshless
   advection using flow-directional local-
   grid method (MPS-MAFL method),
   136—137, 137*f*, 139, 238
MPS for all speed (MPS-AS), 113,
   116—117
MPS method. *See* Moving particle semi-
   implicit method (MPS method)
MPS-AS. *See* MPS for all
   speed (MPS-AS)
mps.c sample program, 74—83
Multiphase flow, 157—158, 189—190, 192,
   201, 229—230
Multiresolutions techniques, 254—258
   dam-break simulation
      by OPT, 258*f*
   droplet deformation simulation, 256*f*
   numerical simulation of rising bubble by
      MPS-MALE, 255*f*

N

Nabla model, 46—47
Nabla operator, 32—33, 45—46
Navier—Stokes equation, 5—7, 16—17,
   31—32, 35—38, 45, 54, 57—58, 61,
   68—69, 93, 119—120, 126—127, 235,
   243, 259, 268
   pressure gradient term, 35—36
   viscous term, 36—38

Negative pressure problem, 200—201, 204
Neighbor list, 5
Neighbor particle list, 168
Neighboring bucket, 252—253
Neighboring particle, 4—5, 50—51,
    54—55, 74, 83—84, 106
Neighboring search, 252, 252f
Neumann boundary condition, 73—74,
    106, 164—166, 171, 191
Newton's second law, 6—7
Newton—Raphson method, 173—174
No-slip boundary condition of stationary
    wall, 74
Non-Newtonian fluid, 265
Non-slip condition, 163, 165—166
Nonhomogeneous Neumann boundary
    condition, 165—168
Nonlinear characteristics of state equation,
    115
Nonuniform particle-size simulations,
    254—255
Nonuniform rational B-spline (NURBS),
    172—174
Normal vector, 196—197, 221—222
Nuclear reactors, 235—236, 235f
Nucleate boiling, 138—139, 138f, 241
Number of space dimensions, 47, 49
Numerical boundary layer, 166
Numerical cavity, 206—207, 211f
Numerical diffusion, 10, 156—157
Numerical stability, 22, 122, 146—147,
    175—176, 205, 222—223, 226, 247
NURBS. See Nonuniform rational B-
    spline (NURBS)

O
Oil
    behavior in commercial gear box,
        263—264, 265f
    circulation in gear box, 260—261, 261f
One-dimension (1D)
    method, 56
    pure convection problem, 7, 7f
One-fluid model, 190—191
Open boundary conditions, 158—159,
    208

OPT. See Overlapping particle technique
    (OPT)
Organs relating to swallowing, 271, 272f
Outflow zone, 210—212, 212f
Outlet boundary, 107, 158—159
    modeling, 207—212, 209f, 211f
Over mirror region, 175—176
Overlapping particle technique (OPT),
    257, 257f, 258f

P
PAF method. See Particle-and-force
    method (PAF method)
Pairwise potential, surface tension
    calculation based on, 223—228
    droplet oscillation, 226f
    improvement of potential-based
        approach, 227
    potential-based model proposed by
        Kondo et al., 224—227, 224f, 225f
    static contact angle calculation, 228f
    wettability calculation, 227—228
Parallel computing, 144—146, 250—254
    buckets for neighboring search, 252f
    domain decomposition technique based
        on buckets, 253f
    tsunami run-up analysis on coastal city,
        253f
ParaView, 85—86
Partial derivative operators, 45
Partial differential operators, 32—33,
    62—63, 71
    approximation of, 45—57
    example of gradient calculation, 51—53
    gradient, 46
    gradient model of MPS method,
        46—47, 48f, 50f
    Laplacian model of MPS method,
        53—57
    Laplacian operator and uses, 53
    meaning of parts of gradient model,
        47—51
Particle accelerations, 31, 95
Particle i. See ith particle
Particle ID, 27—28
Particle interaction models, 12—16, 13f

Particle method, 1−11, 2*f*, 157−159, 208, 217
  continuum mechanics, 5−11
  Lagrangian description, 2−3
  Meshless discretization, 3−5, 4*f*
  MPS method, 11−20
  research history of particle methods, 20−22
  surface tension models in
    applications using MPS method, 229−231
    calculation based on pairwise potential, 223−228
    calculation in MPS algorithm, 218*f*
    calculation using CSF continuum equation, 218−223
Particle number density, 18, 21, 40−45, 70, 73, 95, 112, 116, 181−182, 219−221, 245−247, 254−255
  example of calculation, 44−45
  and fluid density relationship, 43−44
  standard particle number density n^0, 42−43
Particle-and-force method (PAF method), 20−21
Particle-in-cell method (PIC method), 20−21, 239
Particle-mesh hybrid method, 239−240, 239*f*, 242
Particle(s), 2
  approximation, 83−84
  boundary representation, 161−162
  clustering, 201
  deficiency problem, 157−159, 208
  differential equation, 7
  displacement, 59−61
  dynamics, 6−7, 6*f*
  mass of, 33−35
  setting initial positions, 28−29
  setting initial velocities, 29
  shape, 34−35
  shifting technique, 140, 202−203
  simulation, 238
  spatial disorder, 206−207
  type, 86, 89, 107
ParticleNumberDensity[i], 96
Partitioned approaches, 189−191

Passive scalar variable, 7
Periodic boundary conditions, 158−159, 208, 209*f*
Phase change model, 237−238, 237*f*
Phase fraction, 236−237
PIC method. *See* Particle-in-cell method (PIC method)
Plane Poiseuille flow, 208
Plane reflection matrix, 169−171
Poisson equation, 64, 67−68, 97
Polygon boundary representation, 160−162, 180−181
Polygon wall boundary model, 176, 180, 188
Polygon wall model, 176, 188, 189*f*, 190*f*
Polynomial pairwise potential, 230−231
Position vector, 33
Potential force model, 223, 227, 238, 264−265
Potential model
  droplet oscillation using, 226*f*
  wettability calculation in, 227−228
Potential-based approach, further improvement of, 227
Potential-based models, 224−227, 224*f*, 225*f*, 230−231
Potential-based surface tension model, 226−227
Prediction process, 114
Pressure, 35−36, 46−47, 86
  boundary condition of, 73−74
  calculation, 98−99, 200−207, 207*f*
  distribution of particles, 63−64
  field, 202
  fluctuations, 246−247
Pressure gradient term, 35−36, 47, 51−53, 51*f*, 73, 100, 184−185, 240
Pressure Poisson equation, 17−18, 21−22, 38−39, 53−54, 63−64, 69−73, 95, 106, 111−113, 115, 118−119, 123, 131−132, 136−137, 140, 179, 192, 200, 202, 223, 245−246
  calculation, 70−73
  derivation of MPS method, 68−70
  for incompressibility, 111
Pressure term. *See* Pressure gradient term
Pressure wave propagation, 113, 117
Prof file-format, 92

Program contents, 74–84
Pseudo-compressibility, 253–254
  explicit algorithm using, 118–125
Pure convection, 7

Q

Quality management (QM), 260
Quasi-homogeneous particle distribution, 160
Quaternion, 142–143, 143*f*
QUICK scheme, 135–136

R

RATTLE algorithm, 131, 133–134, 147
Ray, 169–171
Re-meshing, 10
Real particles, 204–205
Reference density, 120
removeNagativePressure() function, 96
Renumbering of particles, 252
Reynolds stress, 243
Rigid body model, 140–146, 141*f*
Rotation matrix, 143

S

Saint Venant-Kirchhoff model, 147–148
Saturation temperature, 238
Scalar parameter, 32, 62
Scanning scheme, 195–197, 195*f*
Second partial differentiation, 37–38
Second-order explicit symplectic scheme, 128–129
Semi-implicit algorithm, 16–19, 21–22
Semi-implicit method, 57–74, 108
  boundary condition of pressure, 73–74
  boundary condition of velocity, 74
  derivation of pressure Poisson equation of MPS method, 68–70
  in MPS method, 59–61
  particle motion, 60*f*
  pressure and necessity calculation, 57–59
  pressure Poisson equation calculation, 70–73
Semi-Lagrangian, 136

setBoundaryCondition() function, 95, 97, 97*f*
setMatrix() function, 98–99, 99*f*
setMinimumPressure()function, 96
setSourceTerm() function, 95, 98, 98*f*
Shear rate, 266–267
Shear stress, 265
Shortening computation time, techniques for, 83
Simulation programs, 74–102
  compiling and executing sample programs, 84–85
  contents of program, 74–84
  functions, 75*t*, 86–102
    calculateGravity function, 93
    calculateNumberDensity() function, 96
    calculateNZeroAndLambda() function, 90
    calculatePressure() function, 95–96
    calculatePressure_forExplicitMPS() function, 101–102
    calculatePressureGradient() function, 100–101
    calculatePressure Gradient_forExplicitMPS() function, 102
    calculateViscosity function, 93–94
    initializeParticle PositionAndVelocity_for 2dim() function, 88–89
    libraries and declarations, 86–87
    main() function, 88
    mainLoopOfSimulation() function, 91–92
    moveParticle() function, 95
    setBoundaryCondition() function, 97
    setMatrix() function, 98–99
    setSourceTerm() function, 98
    solveSimultaniousEquations ByGaussianElimination() function, 99
    weight() function, 90–91
  symbolic constants, 77*t*
  variables, 79*t*
  visualization, 85–86
Simultaneous linear equations, 58–59, 63–64, 71–73, 83–84, 97–99, 108

Single-phase free-surface flow, 192
Slight linear compressibility, 111−112, 115
Slightly compressible MPS, 112
Slip boundary condition, 74
SMAC method, 61
Smagorinsky constant, 243−244
Smoothed particle hydrodynamics method
    (SPH method), 5−6, 118−119, 128,
    159, 181
    MPS and, 18−20
Solid fraction, 233−234
Solid wall boundaries, 158−188
    boundary integral−based polygon
        representation, 180−188
    boundary representations for solid wall,
        160f
    distance function−based polygon
        representation, 176−180
    evaluation of discretization, 161f
    mirror particle representation, 168−176
    wall particle representation, 162−168
solveSimultaniousEquations
    ByGaussianElimination() function, 99,
    100f
Sound speed, 11, 22, 28−29, 122
Source codes, 74
Source term, 64, 69−70
    of pressure Poisson equation, 18, 21−22,
    64, 70, 112−113, 136−137, 246
Space, 3
Sparse matrix, 72−73, 84
Spatial resolution, 28, 103−104
Specific heat, 233−234
Speed of sound, 108
SPH method. *See* Smoothed particle
    hydrodynamics method (SPH method)
Spring-based model, 147
SPS turbulence model. *See* Subparticle-
    scale turbulence model (SPS turbulence
    model)
sqrt (mathematical operator), 94
Stabilization techniques, 191
Standard mirroring rule, 174−175
Standard particle number density n^0,
    42−43
Static contact angle calculation using
    potential model, 228, 228f

Stationary walls, 163−164
stdio.h functions, 86−87
Stokes flow, 169−171
Störmer-Verlet scheme, 128−129
Structural analysis, 146−150
Subparticle-scale turbulence model (SPS
    turbulence model), 242, 244
Substantial derivative, 32, 61−62
Suppression of pressure fluctuations,
    245−247
Surface
    deformation, 156−157
    energy, 225
    integration, 142, 182−183
    mesh, 176
    roughness, 165
    shape, 34−35
    tension, 190−191
    tension force, 189−190
        acting at liquid particles, 239, 239f
Surface tension models in particle
    methods. *See also* Continuum surface
    force models (CSF models)
    applications using MPS method,
        229−231
    surface tension calculation
        based on pairwise potential, 223−228
        using CSF continuum equation,
            218−223
    in MPS algorithm, 218f
Swallowing process, 271, 273f
Swallowing simulation, 29
Symplectic Euler scheme, 126
Symplectic scheme, 125−134

T

Taguchi method, 263, 264f
Tait equation, 113, 117−119
Temporal discretization, 8
Temporary velocity, 62, 64−65, 245−246
Temporary velocity vector, 17−18, 115
Thermal energy conservation equations,
    236−237
Three-dimensional space, 140
Time increment (Δt), 30, 87, 106−107
Time integration, 62−63
Time-consuming process, 250−251

Time-dependent diffusion process, 15—16
Total number of particles, 5
Traditional mesh methods, 156
Traditional mesh-/grid-based methods, 159
Transposition symbol, 33
True velocity vector, 64
Tsunami run-up analysis, 124—125, 145*f*, 253*f*
Turbulence kinetic energy, 243
Turbulence model, 242—244
Two dimension (2D)
    dam break problem, 188, 190*f*
    hydrostatic pressure problem, 188, 189*f*
    simulation, 53—54, 87
    space, 140
Two principal curvatures, 173

U

ULSI. *See* Upwind least square interpolation (ULSI)
Un-realistic local particle oscillation/un-realistic particle oscillation, 146—147
Units of variables, 87
UNIX environment, 85
Unknown parameter, 71
Unknown values, number of, 67
Upstream particle, 212, 212*f*
Upwind least square interpolation (ULSI), 250
Upwind weight function, 139—140, 139*f*

V

V&V. *See* Verification and validation (V&V)
Variable declaration, 28
Variable DT, 30
Variational formulation, 181
Vector function, 182—183
Vector notation, 39—40
Velocity, 210—212
    boundary condition, 74
    distribution, 37—38, 65—66
    vector, 243
    on wall, 38

Verification and validation (V&V), 259—260
    application to automobile industry, 260—265
    application to biomechanics, 270—273
    application to chemical engineering, 265—268
    application to metal engineering, 268—270
Virtual boundary particles, 176—177
Virtual particle, 204
Viscosity, 235
    coefficient, 31—32
Viscous force, 36—38, 59—61, 93—94, 164
Viscous term, 36—38
Visibility criterion, 165
Visibility determination, 175—176
VOF method. *See* Volume of fluid method (VOF method)
Volume
    integration, 33, 184—185
    of particle, 34—35
Volume of fluid method (VOF method), 217
Vtu file-format, 91—92

W

Wall contribution term, 162—165, 172—173, 176—178, 188
Wall function, 244
Wall particles, 73—74, 105—106, 164—165, 164*f*
    method, 188
    representation, 162—168
        interpolation points, 167*f*
        wall particles and dummy particles, 164*f*
Wall shear stress, 163—164
Wall weight function, 176, 179
Water
    behavior around car, 261, 262*f*
    column collapse, 26, 26*f*, 83
    droplet impingement on water film, 241—242
    motion, 270—271
    wave propagation, 132—133, 132*f*

WCSPH algorithm. *See* Weakly compressible SPH algorithm (WCSPH algorithm)
Weakly compressible explicit algorithm, 22
Weakly compressible SPH algorithm (WCSPH algorithm), 21–22, 118–119
Weber number, 241
Weight function, 12, 12*f*, 14–15, 34–35, 40–45, 41*f*, 42*f*, 48–49, 54–55, 90–91, 116, 121, 146–148, 182
weight() function, 90–91, 91*f*
Weighted difference, 11–12

Welding process, 236, 269–270
Wettability calculation in potential model, 227–228
Windows environment, 84–85

Y

Young's equation, 227–228

Z

Zero limiter, 200–201, 207*f*

Printed in the United States
By Bookmasters